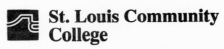

A Color Atlas of

PLANT STRUCTURE

Bryan G. Bowes

Senior Lecturer
Division of Environmental and Evolutionary Biology
University of Glasgow, Scotland, UK

With color drawings by Jo Nicholson

Iowa State University Press/Ames

Published in the United States by
Iowa State University Press
2121 South State Avenue
Ames, Iowa 50014–8300

Library of Congress Cataloging-in-Publication Data

Bowes, Bryan G.
     A color atlas of plant structure/Bryan G. Bowes: with color
drawings by Jo Nicholson
          p.   cm.
     Includes bibliographical references (p. 173) and index.
     ISBN 0–8138–2687–X
     1. Botany—Anatomy—Atlases.   I. Title.
  OK642.B68   1996
  581.4'022'2—dc20

Copyright © 1996 Manson Publishing Ltd
ISBN: 1-874545-20-0 Cased edition

Edited, typeset and designed by Paul Bennett, Tonbridge, Kent, England.
Color separations by Tenon & Polert, Hong Kong.
Printed by Grafos S.A., Barcelona, Spain.

Contents

Dedication

To my parents, especially my mother, for always encouraging my studies. To my late wife, Ruth, for her support in my earlier university career. To my wife, Diane, for her warm but critical interest, general enthusiastic support and forbearance. Also to my children, Tanya and Adrian, for being all that they are.

How can you buy or sell the sky, the warmth of the land? The idea is strange to us. If we do not own the freshness of the air and the sparkle of the water, how can you buy them?

Every part of this earth is sacred to my people. Every shining pine needle, every sandy shore, every mist in the dark woods, every clearing and humming insect is holy in the memory and experience of my people.

From *The Great Chief Sends Word: Chief Seathl's Testament.*

Acknowledgements

I wish to give very warm appreciation to several people for their invaluable assistance in various aspects of the preparation of this Atlas: Mrs Jo Nicholson, for the skilled and artistic production of the numerous colour diagrams in the text; Mr T. Norman Tait for photographic advice and assistance, printing of electronmicrographs and the loan of a number of illustrations (as acknowledged in the figure legends); and Mrs Pamela McEwan for her skill and great dedication in preparing the manuscript from my rough and sometimes almost illegible handwritten copy. I should also like to thank Dr James Dickson for the loan of two electron micrographs of pollen, as acknowledged in the figure legends. I also wish to extend my gratitude to the numerous technicians who, over the last 30 years, have prepared for me many of the specimens illustrated. Finally, I should like to acknowledge the use of various slides and specimens from the Botany Department collection at Glasgow University used to illustrate this Atlas.

Abbreviations

ER	endoplasmic reticulum
G-Os	fixation in glutaraldehyde followed by osmic acid
Km	fixation in potassium permanganate
LM	light microscopy
LS	longitudinal section
μm	micrometre; one thousandth of a millimetre
nm	nanometre; one thousandth of a micrometre
Phase LM	phase contrast light microscopy
Polarised LM	polarised light microscopy
RLS	radial longitudinal section
RER	rough endoplasmic reticulum
SEM	scanning electron microscopy
TEM	transmission electron microscopy
TS	transverse section

Preface

'Why should they care about the histogenesis of the leaf, or adventitious roots? ... The public wants heart transplants, a cure for AIDS, reversals of senility. It doesn't care a hoot for plant structures, and why should it? Sure it can tolerate the people who study them ... They're relatively inexpensive too. It costs more to keep two convicts in Statesville than one botanist in his chair.'

More Die by Heartbreak, Saul Bellow

While Bellow's character Kenneth Trachtenberg may convincingly relegate the study of plant structure to a backwater, it is a commonplace that we are all ultimately dependent on green plants for our survival on earth. Horticultural successes in increasing crop yields and developing new plant varieties emphasise the importance of plant physiology, biochemistry and molecular biology, for all of which the study of the green plant's internal form and internal structure is a prerequisite.

With the steadily increasing content of undergraduate and graduate biology courses, the proportion of a student's time devoted to plant morphology is inevitably reduced. There is no longer time for detailed study of the excellent and exhaustive texts in plant anatomy, and many find that plant structures are most easily understood when mainly described by annotated photographs and drawings.

Such is the concept behind the present atlas: knowledge of plant structure is fundamental to the study of plant science, and that knowledge has to be imparted clearly, briefly and precisely.

Following an introductory chapter on the morphology of the vascular plant, there are seven chapters each dealing with a major aspect of plant structure. A comprehensive glossary of botanical terms used in the atlas is also included. The text for each chapter sets out the essential characteristics of the plant features described and makes extensive reference to appropriate illustrations in the particular chapter and elsewhere in the book. Each illustration is accompanied by a legend and salient features are numbered (not labelled) for maximum clarity, referencing the structures to a boxed key. It is hoped that these aspects of the atlas, together with the photographs and drawings, will prove attractive and useful to many readers.

The atlas is intended for use in different ways by different readers. For the university or college student, the atlas is intended to be read either as a concise introductory text or as a revision guide in preparation for exams. For the professional instructor or for the researcher in academic life or in industry, it is hoped the atlas will provide a source of rapid reference. For the artist or the amateur student of natural history, the intrinsic beauty of many plant specimens, in external form and under the microscope, is clearly shown in the photographs and drawings, arranged in sequence after the text in each chapter.

The overall intention has been to provide a concise and highly illustrated summary of present knowledge of the structure of vascular plants, with particular emphasis on flowering plants.

Bryan Bowes
Bearsden, Glasgow
September 1995

CHAPTER 1

Introduction

The assortment of land plants

Flowering plants, or angiosperms (**1.1**), dominate large areas of the land surface and represent the climax of vascular plant evolution. They occupy a wide range of habitats and about a quarter of a million species have been recognised so far. However, many more, particularly from tropical regions, await scientific description. Angiosperms are very diversified in their form and range in size from a few millimetres in diameter in the aquatic *Lemna* to 90 metres or more in height in *Eucalyptus* (**1.2**).

Flowering plants provide the vast majority of those eaten by man or utilised for animal fodder. Likewise, angiosperms provide various hardwood timbers (**1.2**, **1.3**), many of which are important commercially (*Acacia, Carya, Eucalyptyus, Fagus, Quercus*), fibres (*Corchorus, Linum,* **1.4**) and drugs (*Coffea, Papaver*). Most decorative garden plants are grown to provide floral displays (**1.1**).

Although the flowering plants are now dominant in many habitats, remnants of vascular plant groups which evolved earlier are still present in the flora. There are about 600 species of gymnosperms (conifers some of which provide various softwood timbers, cycads and others) whose seeds are naked (**1.5**); this distinguishes them from the flowering plants where the seeds are enclosed within a fruit (**1.6**). The conifers dominate the vast tracts of boreal forest which occur in North America and northern Europe and Asia. The spore-bearing ferns and their allies (**1.7**, **1.8**) number about 10,000 species.

As well as over 260,000 species of vascular plants, the land flora includes the non-vascular, spore-bearing bryophytes (**1.9**, **1.10**). These small plants comprise about 20,000 species of moss (**1.10**) but far fewer liverworts (**1.9**) and hornworts. They lack cuticular covering to the epidermis and are usually confined to moist locations. Hornworts and liverworts are often simply-organised thalloid structures without leaves (**1.9**), but mosses (and some liverworts) are more complex and have leafy green shoots (**1.10**).

The stems of many mosses and a few liverworts show a central strand of tissue, apparently concerned with the movement of water and soluble foodstuffs. However, except in a few taxa (e.g. *Polytrichum*), this does not have the structural complexity of the xylem and phloem tissues (**1.11**) of vascular plants. The latter tissues are concerned with the rapid, long-distance transport of water and soluble foodstuffs (**1.12**).

Lichens are not true plants but rather symbiotic associations of fungi and algae; these however, often show a complex plant-like form (**1.13**).

Transpiration and translocation in vascular plants

The root system absorbs water, together with dissolved mineral salts, from the soil. This passes across the cortex and endodermis of the young root to the central xylem (**1.14**). The dead tracheary elements of this tissue have strong thickened walls (**1.11**) and their lumina are filled with columns of water moving upwards into the shoot (**1.12**).

This transpiration stream is powered by the evaporation of water vapour from the shoot surface, and mainly occurs through the stomata in the leaf epidermis. These small pores (**1.15**) normally remain open in the day and allow the entry of carbon dioxide, which is essential for photosynthesis in the green foliage. The sugars thus elaborated are translocated (**1.12**) in solution in the living sieve elements of the phloem (**1.11**) to the stem and root where they are either stored (**1.14**) or metabolised.

General morphology of angiosperms

The young shoot of the generalised dicotyledonous flowering plant (**1.16**) bears a number of leaves and normally a lateral bud occurs in the axil of each leaf. The leaf is attached to the stem at the node, while the internode lies between successive leaves. The leaf is usually flattened and often bears a leaf stalk (petiole, **1.17**). In a horizontal leaf the adaxial surface (which was nearest to the shoot apex while within the bud, **1.18**) lies uppermost, and the abaxial side forms the lower surface.

A simple leaf may be dissected or lobed, and a compound leaf shows several leaflets (**1.19**); these, however, do not subtend axillary buds. In the lamina (leaf blade) a network of veins is present (**1.20**) which links to the vascular system of the stem. The axillary (lateral) buds may remain dormant but normally develop into side shoots, or form flowers. At the base of the main stem the cotyledons (first leaves formed in the embryo) demarcate it from the hypocotyl; the latter represents a transition zone between stem and root.

In the angiosperms two groups have evolved which show distinct morphological, anatomical and floral characteristics (**1.21**). The dicotyledons (crucifers, begonias, willows, oaks) constitute about two-thirds of flowering plant species and the great majority show some degree of secondary (woody) thickening (**1.2** to **1.4**). Monocotyledons (grasses, bananas, lilies, **1.1**) do not undergo secondary thickening in the same way as dicotyledons, but in some genera large trees may develop (**1.22**).

Dicotyledonous leaves are commonly petiolate (**1.17**) and normally show a narrow attachment to the stem. In monocotyledons the leaf is frequently sessile (without a petiole) and the leaf base often encloses a large sector of the stem (**1.23**). Leaves of dicotyledons are varied in shape and arrangement of their major veins but normally show a reticulate pattern of the small interconnecting veins (**1.20**). In monocotyledons the leaf is typically elongate (**1.22**), with the main veins paralleling its length. Their relatively rare lateral connections are normally unbranched (**1.24**).

In dicotyledons, the radicle (seedling root) is normally persistent and the older region often increases in diameter by secondary thickening (**1.25**). By contrast, in monocotyledons the radicle is usually not persistent and an adventitious root system develops from the base of the enlarging shoot. In a number of larger monocotyledons their heavy upright shoots are stabilised by adventitious proproots (**1.26**).

Vascular anatomy of angiosperms

The primary vascular systems of mono- and dicotyledons generally differ considerably (**1.21**). In a transverse section of the monocotyledonous stem (**1.27**) there are many scattered vascular bundles, while in dicotyledons a smaller number of bundles is usually arranged in a cylinder outside a wide pith (**1.28**). The roots of monocotyledons frequently show a central pith with a large number of strands of alternating xylem and phloem on its periphery (**1.29**). But in the dicotyledons a star-shaped core of xylem commonly occurs with strands of phloem lying between its several arms (**1.14**).

In the majority of dicotyledons a fascicular cambium separates the primary xylem and phloem of both the stem and root (**1.25, 1.28, 1.30**). If secondary thickening occurs the normally discrete strands of cambium become linked, and the continuous ring of vascular cambium produces secondary xylem internally and secondary phloem externally (**1.4, 1.30**). The vast majority of monocotyledons are herbaceous; however, a number of palms grow into tall trees as a result of diffuse secondary growth. Others (*Dracaena, Cordyline*) produce new (secondary) vascular bundles from a secondary thickening meristem and may form large trees (**1.22**).

Floral and reproductive features of angiosperms

In monocotyledons the floral parts (sepals, petals, stamens and carpels) commonly develop in threes (**1.1, 1.21**), whereas in dicotyledons these frequently occur in fives or fours (**1.21**). However, a large and indefinite number of floral parts occur in many other dicotyledons (**1.31**). The mature carpel (female part of the flower, **1.31, 1.32**) consists of several parts: the terminal stigma which receives the pollen (**1.1, 1.31**), an intermediate style (**1.1**) and the basal ovary (**1.32**). In most taxa the carpels are fused (syncarpy, **1.6**) rather than free from each other (apocarpy, **1.31**). Within the ovary, one to numerous ovules are present and each contains an egg cell at the micropylar end of the ovule (**1.32**). The pollen grain germinates on the stigma and the pollen tube grows down the style to enter the ovule (**1.32**) where it liberates two haploid sperm nuclei.

One of these fertilises the egg which forms the diploid zygote, while the other nucleus fuses with the two centrally located polar nuclei (**1.32**) to give rise to the nutritive endosperm for the embryo. As the embryo develops from the zygote it enlarges and the surrounding tissues of the ovule expand to form the mature seed. The ovary concomitantly increases in size to form the mature fruit (**1.6**). In dicotyledons two cotyledons are present on the embryo (**1.21, 1.33**), but in monocotyledons only a single one occurs (**1.21**).

Theme of the atlas

This book is concerned with the development and mature form of the vascular plant and attention is focused on its structure at an anatomical, histological and fine-structural level. As previously emphasised, the angiosperms represent the present climax of plant evolution and dominate a varied range of habitats. They are the most numerous members of the land flora and provide nearly all of the plants, except for conifers which yield softwoods, exploited economically by man. It is therefore appropriate that the examples discussed in ensuing chapters concentrate on the varying manifestations of the anatomy of the flowering plant on which we are all so dependent.

This Atlas is intended to serve both as an integrated series of clearly described illustrations and a concise text. It is believed that this presentation will facilitate the appreciation and understanding of basic plant anatomy. It is intended for use by biology undergraduates, as well as by graduates who have no previous knowledge of plant structure but are undertaking research in plant physiology, biochemistry, horticulture or related fields. The Atlas will also be relevant to biological studies at advanced school and college levels, whilst the abundant and fully annotated illustrations should be of general interest to a wider audience.

A short bibliography details various texts which supplement the essentials of plant anatomy presented in this work, whilst the glossary will be of especial use to the reader where illustrations and legends are consulted without direct reference to the text.

1.1 Hermaphrodite flowers of *Lilium* 'Destiny' (lily). The floral parts are grouped in threes: six orange-spotted yellow perianth members, six stamens with orange anthers (1) and a three-lobed orange stigma (2). Floral parts in threes typify monocotyledons as do also elongate, narrow leaves (cf., **1.21A**).

1	Anthers
2	Stigma

1.1

1.2 Gigantic trunk of the dicotyledon *Eucalyptus diversicolour* (Karri tree). This is a native of S.W. Australia and specimens up to 87 metres tall and with a diameter of 4 metres have been recorded. These trees have been extensively logged for their very strong timber which is used in the construction industry.

1.2

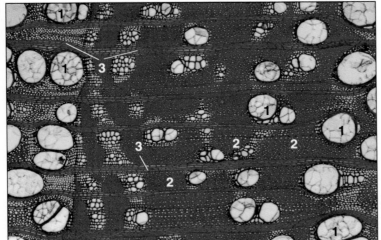

1.3 TS illustrating ring porous wood (secondary xylem) of *Robinia pseudoacacia*, a dicotyledon. Large pores (vessels) with thick secondary walls are confined to wood formed early in the growing season and their lumina are packed with tyloses (1). Numerous small vessels are clustered in the later wood and there is an abrupt boundary between the late wood of one year and the early wood of next year.

The vessels are embedded in a mass of narrow, very thick-walled fibres (2). The axial parenchyma consists of narrow-diametered thin-walled cells associated with the vessels. Numerous parenchymatous rays (3), several cells wide and with thickened walls, traverse the wood radially. (LM x 35.)

1	Tyloses
2	Thick-walled fibres
3	Parenchymatous rays

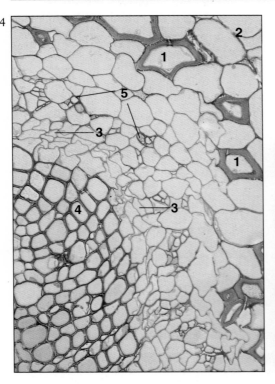

1.4 TS of the stem of the dicotyledon *Linum usitatissimum* (flax) showing numerous large and thick-walled fibres (1) which are maturing adjacent to the thin-walled cortical parenchyma (2). Within the vascular cylinder a narrow cambial layer (3) is giving rise to secondary xylem (4) while groups of narrow-diametered elements (5) represent the translocating elements of the phloem. (G-Os, LM x 280.)

1	Thick-walled fibres
2	Cortical parenchyma
3	Cambial layer
4	Secondary xylem
5	Translocating phloem

1.5 Mature tree of the gymnosperm *Cycas circinalis*, bearing numerous freely-exposed seeds. Unlike angiosperms, these are not enclosed within a fruit (cf., **1.32**). Note the very large and compound vegetative leaves. (*Copyright of T. Norman Tait.*)

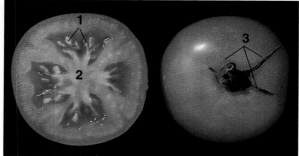

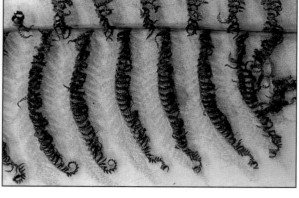

1.6 *Lycopersicon esculentum* (tomato) fruit. In TS the numerous seeds (1) are seen attached to axial placentas (2) and embedded in juicy tissue derived from placental tissue. The four locules indicate that the fruit developed from an ovary of four fused carpels. In the external view the remains of the flower stalk and the green sepals (3) are visible at its base; this fleshy fruit developed from a single superior ovary which was situated above the perianth.

1	Seeds
2	Axial placentas
3	Sepals

1.7 Portion of a withered fertile frond (leaf) of the subtropical tree fern *Dicksonia antartica*. The multitudinous sporangia on its abaxial surface have dehisced and shed billions of yellow, uniform-sized, haploid spores onto the bench surface.

1.8 LS of the heterosporous cone of *Selaginella* (a fern ally). The sporangia are borne in the angle between the cone axis and the sporophylls (1). Within the sporangia the spore mother cells undergo meiosis: the megasporangia contain a few large and thick-walled haploid megaspores (2), while the microsporangia contain abundant small and thin-walled microspores (3).

Note the small ligules (4) on the adaxial surface of the sporophylls; this feature shows the affinity of present day *Selaginella* with the fossil *Lepidodendron*, the dominant tree of the Carboniferous coal measures. (LM x 35.)

1	Sporophyll
2	Megaspores
3	Microspores
4	Ligule

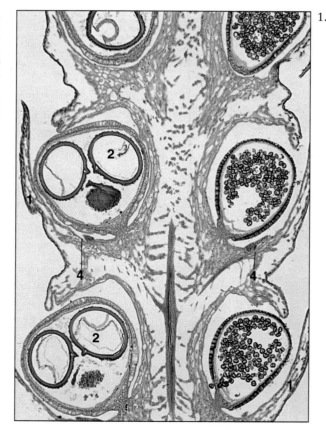

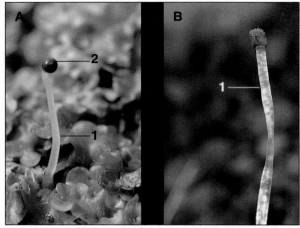

1.9 **A**: closely crowded gametophytic plants of the liverwort *Pellia epiphylla*. Each green, flattened and lobed thallus bears male and female sex organs and after fertilisation the embryo develops into the diploid sporophyte dependent on the gametophyte. A single mature sporophyte is visible, consisting of a seta (1) and terminal capsule (2). **B** shows a dehisced capsule with some yellow spores still evident.

1.10 Closely crowded gametophytes of the moss *Mnium hornum*. The unbranched leafy stems grow in tufts up to five centimetres high in shaded woodlands; the specimens shown were attached to a piece of tree bark. The leafy shoots bear male or female sex organs and, although longer-lived and more complex than liverworts, the moss sporophytes are also dependent on the leafy gametophytes.

1	Seta
2	Capsule

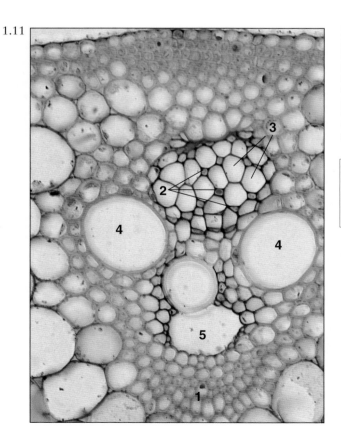

1.11 TS of the stem of the monocotyledon *Zea mays* (maize) showing a peripheral vascular bundle. The bundle is invested by many small, thick-walled sclerenchyma fibres (1). The phloem lies exarchly and its smaller, densely staining, companion cells (2) contrast with the larger sieve elements (3). The endarch xylem consists of several large-diametered vessels (4) with thickened secondary walls. During internode elongation the innermost element has been overstretched, and its primary wall torn, and is now represented by a cavity (5). (LM x 340.)

1	Sclerenchyma fibres
2	Phloem companion cells
3	Sieve elements
4	Xylem vessels
5	Cavity in xylem

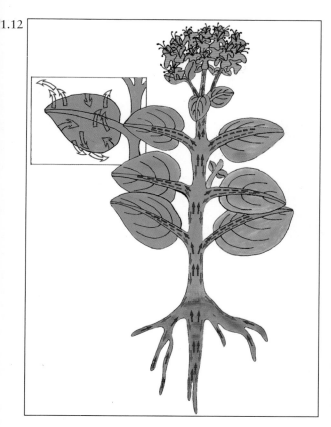

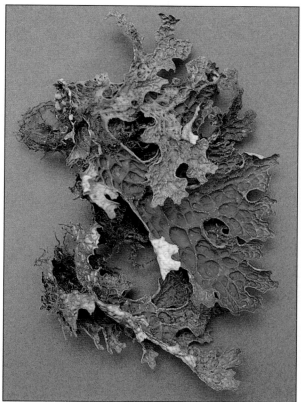

1.12 Diagram of a herbaceous flowering plant illustrating the following functions: I. The flow of water in the xylem (blue arrows) from the root to stem and transpiring leaves. II. The translocation of sugars in the phloem (red arrows) from the photosynthetic leaves to actively-growing organs or storage regions in the shoot and root. III. The evaporation of water vapour (open arrows) into the atmosphere and the diffusion of carbon dioxide (stippled, open arrows) into the leaf via the open stomatal pores.

1.13 The broad lobed 'leafy' lichen *Lobaria pulmonaria* (tree lungwort). This is not a true plant but is composed of several symbiotic partners; embedded within a compact covering of fungal (ascomycete) hyphae lie photosynthetic green algal cells. In this lichen scattered colonies of prokaryotic blue-green algae also occur.

1.14 TS of the mature primary root of the dicotyledon *Ranunculus* (buttercup). Note the parenchymatous cortex (1) packed with red-stained grains of starch and the central vascular cylinder which is surrounded by the endodermis (2). The thick-walled, dead xylem elements (3) conduct water through their lumina to the aerial shoot system. Strands of phloem (4) lie between the xylem arms. (LM x 110.)

1	Parenchymatous cortex
2	Endodermis
3	Xylem elements
4	Phloem

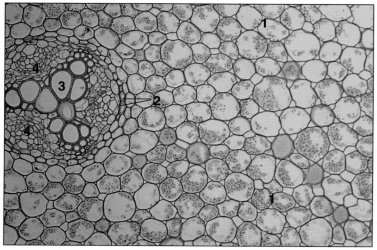

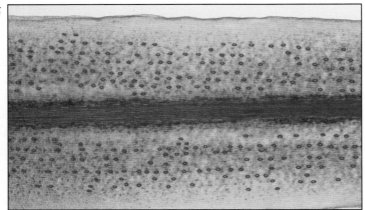

1.15 Cleared leaf of the gymnosperm *Taxus baccata* (yew). This is viewed from the abaxial surface and shows the numerous epidermal stomata orientated parallel to the long axis of the leaf. Note also the single, longitudinal vein which is visible internally. (LM x 25.)

1.16

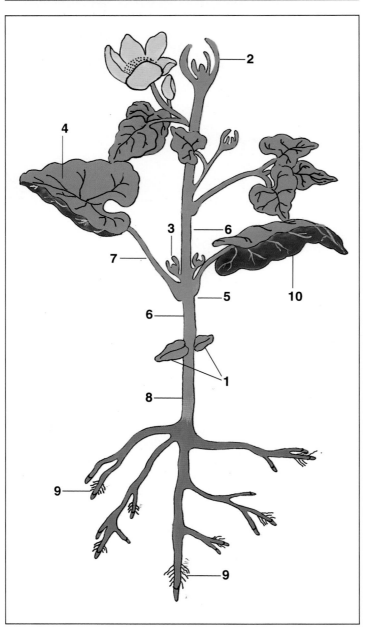

1.16 Diagram showing the morphology of a generalised herbaceous flowering plant. The two small cotyledons (1) at the base of the stem were the first leaves developed on the embryo and indicate that this is a dicotyledon (cf., **1.21B**). The shoot bears a terminal bud (2) whilst a number of lateral buds (3) lie in the axils between the stem and the adaxial face (4) of the leaves. The uppermost bud has developed into a flower, but the remainder are vegetative.

The leaves arise at the nodes (5) whilst the intervening regions of stem are the internodes (6). The leaf consists of the lamina borne on a petiole (7). In the hypocotyl (8) the vascular tissue changes from the arrangement in the stem to that of the root (cf., **1.21**). The tap root and laterals bear numerous root hairs (9). Abaxial surface of leaf (10).

1 Cotyledons	6 Internode
2 Terminal bud	7 Petiole
3 Lateral bud	8 Hypocotyl
4 Adaxial leaf face	9 Root hairs
5 Node	10 Abaxial leaf surface

1.17 Simple leaves of the dicotyledons *Kalanchoe diagromontianum* and *Begonia metallica*. Both are differentiated into lamina (1) and petiole (2). *Kalanchoe* (**A**) shows a single midrib in the lamina but in *Begonia* (**B**) several main veins are arranged palmately. Note the adventitious buds (3) which develop from the notched margins of the *Kalanchoe* blade; when these fall onto the soil they root and develop into new plants.

1	Lamina
2	Petiole
3	Adventitious buds

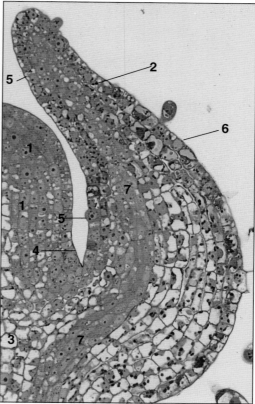

1.18 Median LS through the shoot tip of the dicotyledon *Glechoma hederacea* (ground ivy). Note the hemispherical shoot apex (1) and leaf primordium (2) arising at its base. The core of the apex is somewhat vacuolated and will give rise to the pith (3) of the young stem. An axillary bud primordium (4) is developing at the adaxial base (5) of the leaf. The abaxial face (6) of the leaf shows considerable vacuolation and this is confluent with the cortex of the young stem. Procambium (7). (LM x 325.)

1	Shoot apex
2	Leaf primordium
3	Pith
4	Axillary bud primordium
5	Adaxial leaf face
6	Abaxial leaf face
7	Procambium

1.19 Compound leaf of the dicotyledon *Rosa* (rose). Note the three leaflets, each with a pinnate (feather-like) arrangement of the veins, borne on a petiole with a pair of green stipules (1) situated at its base.

1	Stipules

1.20

1.20 Abaxial surface of the leaf blade of the dicotyledon *Begonia rex*. Note the numerous lateral connections between the palmately arranged main veins. The smallest veins form a reticulum of polygons enclosing small islands of lamina.

1.21

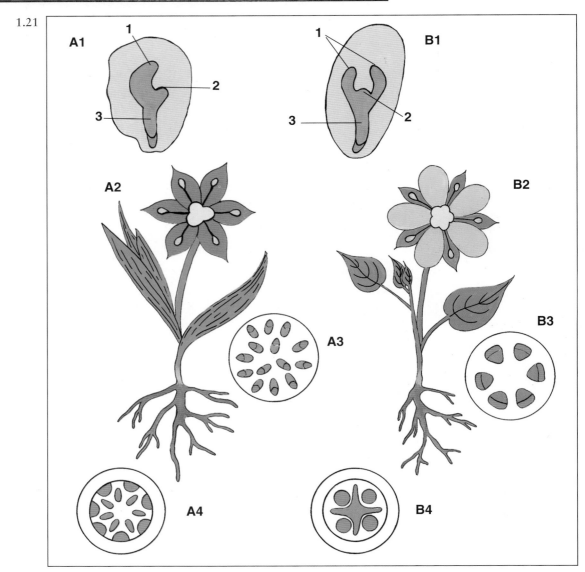

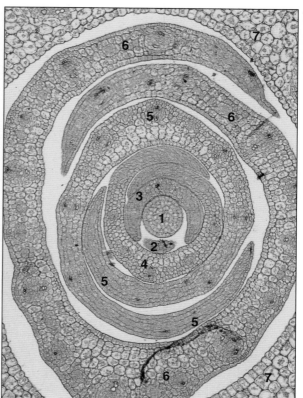

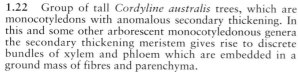

1.22   Group of tall *Cordyline australis* trees, which are monocotyledons with anomalous secondary thickening. In this and some other arborescent monocotyledonous genera the secondary thickening meristem gives rise to discrete bundles of xylem and phloem which are embedded in a ground mass of fibres and parenchyma.

1.23   TS through the seedling shoot apex of the monocotyledon *Zea mays* (maize). The shoot apex (1) is invested by progressively older leaf primordia (2–6) which are arranged in two alternate rows. These leaves are enclosed by the single coleoptile (7), a leaf-like cylindrical structure which protects the plumule before its emergence at germination. (LM x 85.)

1	Shoot apex
2–6	Leaf primordia
7	Coleoptile

1.21   (Left) Diagrams showing the typical differences between monocotyledons (series **A**) and herbaceous dicotyledons (series **B**). **A1** to **B1** are longitudinal sections of seeds; note the single cotyledon (1) in **A1** but paired cotyledons (1) in **B1**. **A2** to **B2** illustrate the following morphological differences: elongate leaves in monocotyledons which are usually without petioles and with parallel veins; in dicotyledons relatively shorter, usually petiolate, leaves occur which show branched and non-parallel veins. The floral parts are in threes in monocotyledons contrasting with fives (or fours or indefinite numbers) in dicotyledons.

A3 to B3 and A4 to B4 show the vascular systems in TS of the stem and root respectively, with xylem coloured blue and the phloem red. The monocotyledonous stem (**A3**) has many scattered vascular bundles consisting of inner xylem and outer phloem, whereas in dicotyledons (**B3**) the stem bundles are usually arranged in a ring with vascular cambium present between the phloem and xylem. The roots of monocotyledons (**A4**) are polyarch with numerous separate strands of phloem separated by xylem and a central pith is usually present. In dicotyledons (**B4**) a pith is often lacking and from the xylem core several arms radiate outwards between the phloem strands. Plumule (2), radicle (3).

1	Cotyledon
2	Plumule
3	Radicle

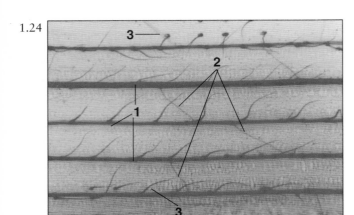

1.24　Cleared leaf blade of the grass *Bromus* (a monocotyledon). The main longitudinal veins (1) are intermittently connected by fine oblique veins (2). Note the numerous trichomes (3) which arise from the epidermis above the main veins. (LM x 6.)

1	Longitudinal veins
2	Oblique veins
3	Trichomes

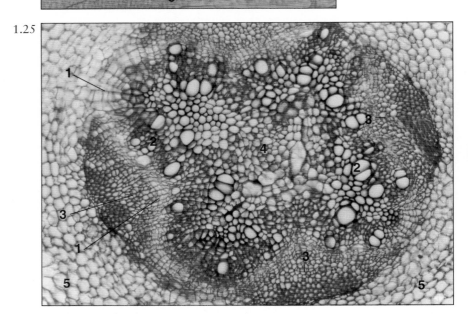

1.25　TS of the older root of *Vicia faba* (broad bean), a dicotyledon. At this distance behind the apex secondary thickening has begun and the convoluted vascular cambium (1) has formed considerable secondary xylem (2) and some secondary phloem (3). Primary xylem (4), cortex (5). (LM x 80.)

1	Vascular cambium
2	Secondary xylem
3	Secondary phloem
4	Primary xylem
5	Cortex

1.26　Trunks of several specimens of the aborescent monocotyledon *Pandanus*. Note the numerous wide-diametered adventitious prop roots which run obliquely outwards from the trunks and grow down into the soil to stabilise the trees.

1.27 TS of the stem of the monocotyledon *Zea mays* (maize).
Numerous scattered collateral
vascular bundles lie in parenchymatous ground tissue and in
each the phloem (1) is situated
nearest to the epidermis (2). Vascular cambium is absent between
the xylem (3) and phloem (cf.,
1.11). (LM x 85.)

1	Phloem
2	Epidermis
3	Xylem

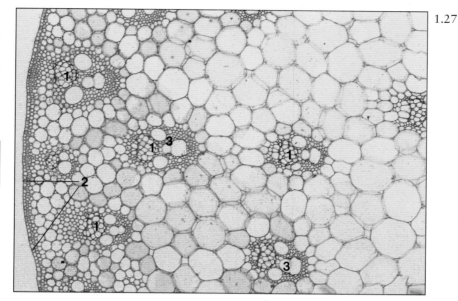

1.28 TS through the young stem of the dicotyledon
Helianthus (sunflower). The collateral vascular bundles lie
in a peripheral cylinder with an extensive pith (1) situated
internally and narrow cortex (2) lying externally. Each
bundle is demarcated on the outside by a cap of phloem
fibres (3) and a prominent fascicular vascular cambium (4)
separates the xylem (5) from the phloem (6); but no
interfascicular cambium has yet differentiated in the
parenchyma between the bundles. (LM x 30.)

1	Pith
2	Cortex
3	Phloem fibres
4	Fascicular vascular cambium
5	Xylem
6	Phloem

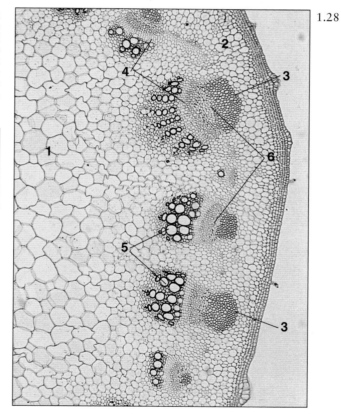

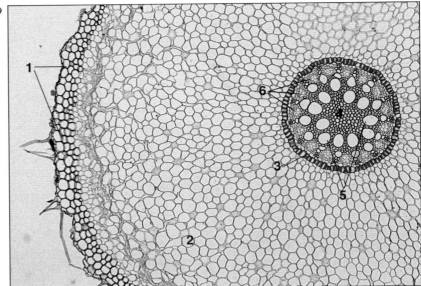

1.29 TS of the old root of the monocotyledon *Iris*. Note the several-layered, thickened, exodermis (1), enclosing the wide parenchymatous cortex (2). A single-layered, thickened endodermis (3) is also present and the vascular tissue enclosed within it shows a polyarch arrangement. The lignified pith (4) is surrounded by radially aligned, wide-diametered xylem vessels (5) but with smaller tracheary elements adjacent to the endodermis, while phloem strands (6) occur between these smaller xylem elements. (LM x 30.)

1	Exodermis
2	Parenchymatous cortex
3	Endodermis
4	Pith
5	Xylem vessels
6	Phloem

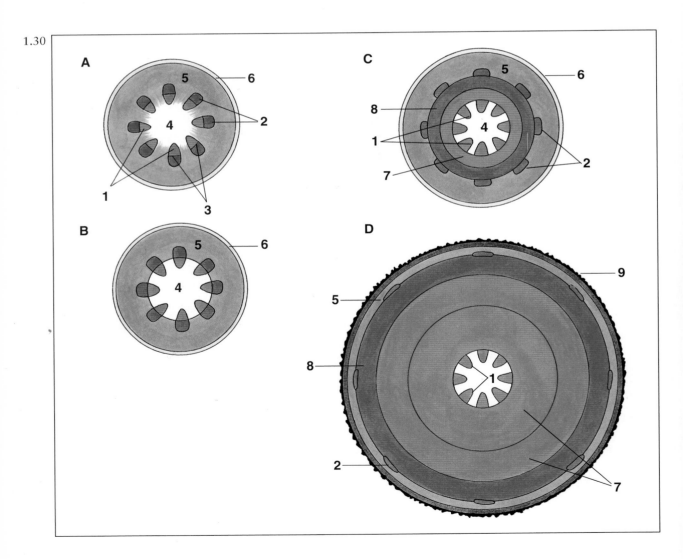

1.31 Inside view of a *Magnolia* flower (a dicotyledon). Some perianth members were removed to show the elongated receptacle which bears numerous spirally-arranged carpels (1) at its tip and numerous stamens below (2). The stigmatic surfaces of the carpels occur on the curved adaxial faces of the styles. The anther occupies the terminal two-thirds of each stamen.

1	Carpels
2	Stamens

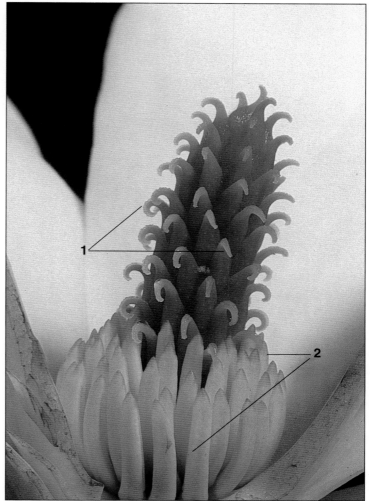

1.30 (Left) Diagrammatic representation of secondary thickening in a dicotyledonous stem in TS. In the young internode (**A**) a ring of vascular bundles occurs with primary xylem (1) lying internally, primary phloem externally (2) and a thin layer of fascicular vascular cambium (3) between them. The bundles are separated by ray parenchyma which extends from the pith (4) to the cortex (5). The epidermis (6) delimits the stem externally.

In **B**, the vascular cambium has become a continuous cylinder and by the end of the first season's growth (**C**) it has formed a continuous ring of secondary xylem (7) internally and a thinner layer of secondary phloem (8) externally. At the end of the next season's growth (**D**), two growth rings are visible in the secondary xylem but the primary phloem and the first year's growth of secondary phloem is becoming disorganised by the expansion of the stem. The epidermis has been replaced by a layer of cork (9) which usually arises in the outer cortex.

1	Primary xylem	5	Cortex
2	Primary phloem	6	Epidermis
3	Fascicular vascular	7	Secondary xylem
	cambium	8	Secondry phloem
4	Pith	9	Cork

1.32

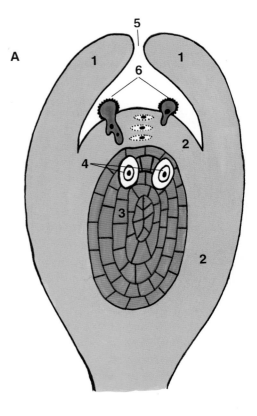

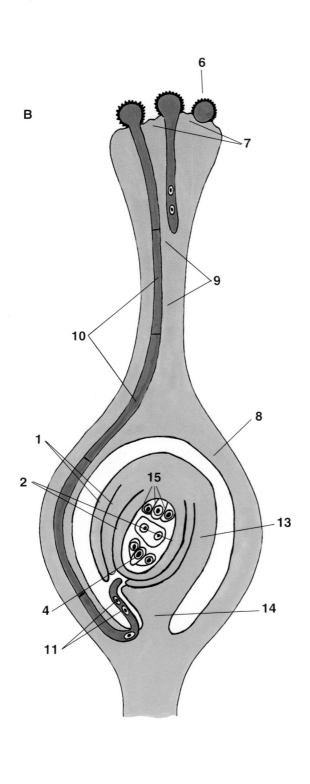

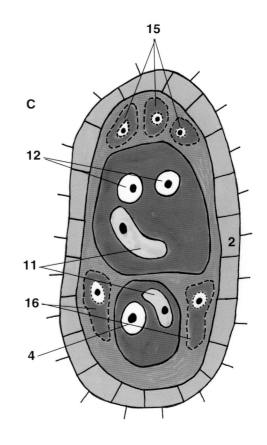

1.32 (Left) Diagrams of sexual reproduction in seed plants. In the naked ovule of a gymnosperm (**A**), an integument (1) invests the megasporangium (2); within, a single haploid megaspore has enlarged and divided to form the female gametophyte (3) which shows several large egg cells (4). The megaspore is one of four meiotic derivatives but the other three (dotted) degenerate. The megasporangium connects with the outside of the ovule via a narrow micropyle (5) allowing entry of haploid pollen grains (6); these have now germinated on the megasporangium surface. The fertilised egg gives rise to a diploid embryo which is nourished by the female gametophyte.

B shows the ovule of an angiosperm within a carpel. The latter comprises a receptive stigma (7) on which several pollen grains (6) have been deposited; a basal ovary (8) and the intermediately-located style (9). The single ovule within the ovary is composed of two integuments (1) which enclose the nucellus (megasporangium) containing a large embryo sac derived from the megaspore. Within the mature sac eight haploid nuclei typically occur, with the egg (lying between a pair of synergids) near to the micropyle. A long pollen tube (10, plugged at intervals by callose) has grown down the style and its tip (adjacent to the micropyle) contains two haploid male nuclei.

C shows detail of the embryo sac of an angiosperm at fetilisation. Two male gametes have been liberated into the embryo sac and one male nucleus (11) is about to fertilise the egg nucleus (4) to form a diploid zygote, while the other will fuse with the two polar nuclei (12) to form triploid nutritive endosperm tissue. The other nuclei in the embryo sac normally degenerate. Funiculus (13), placenta (14), antipodals (15), synergids (16).

1	Integument	9	Style
2	Megasporangium	10	Pollen tube
3	Female gametophyte	11	Male nuclei
		12	Polar nuclei
4	Egg cell	13	Funiculus
5	Micropyle	14	Placenta
6	Pollen grains	15	Antipodals
7	Stigma	16	Synergids
8	Ovary		

1.33 LS of an immature seed of the dicotyledon *Capsella bursa-pastoris* (shepherd's purse). The apical pole of the enclosed embryo shows a rudimentary plumule (1) lying between a pair of prominent cotyledons (2), while the basal pole is terminated by the radicle (3). This is attached to a filamentous suspensor (4) which terminates in a large basal cell (5) at the micropyle (6). The embryo is enclosed by the nucellus (7) and integuments (8); within the embryo sac cellular endosperm (9) is forming. (LM x 125.)

1	Rudimentary plumule
2	Cotyledons
3	Radicle
4	Filamentous suspensor
5	Basal cell
6	Micropyle
7	Nucellus
8	Integuments
9	Cellular endosperm

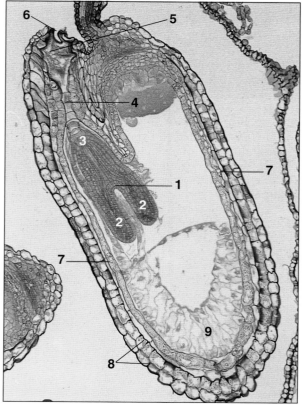

1.33

CHAPTER 2

The plant cell

Introduction

Even in a small green plant there are millions of cells; the vast majority of these are differentiated and in the vascular plant often perform specialised functions such as the transportation of water and soluble nutrients in the xylem and phloem (**1.11**, **1.12**). All differentiated cells originate from actively dividing meristematic cells (**2.1**); these are densely cytoplasmic (**2.2**) and are located in the apical (**1.18**) and lateral meristems (**1.4**, **2.3**) as well as other more localised regions (e.g. meristemoids). Although most living cells are uninucleate (**2.4**) several or many nuclei may occur in certain types (**2.5**), while sieve elements (**1.11**) contain degraded protoplasts (**2.6**) in which the nucleus and most other organelles have senesced.

During the differentiation of most sclerenchyma and tracheary elements their protoplasts also degenerate (**2.7** to **2.9**) and at maturity only their thickened walls remain (**1.11**, **2.10**). The lumina of such dead cells (**2.11**) form a significant part of the plant apoplast system while the walls and intercellular spaces constitute the remainder. The protoplasts of the living cells form the symplast (**2.11**) and these protoplasts are in continuity with each other via the plasmodesmata (**2.11** to **2.13**). Between 1,000 and 10,000 such protoplasmic connections occur per cell, but it seems that only relatively small molecules (with less than a molecular weight of 1,000) are able to pass through them.

The fine-structure of a partly differentiated cell is illustrated diagramatically in **2.14**. The external primary wall is delimited from the protoplast by the membranous plasmalemma. Several large organelles are present including the nucleus, several vacuoles and a number of chloroplasts. Normally under the light microscope (LM), only the wall and these larger organelles would be apparent (**2.1**, **2.15**). However, with the greater resolution of the transmission electron microscope (TEM), mitochondria, endoplasmic reticulum, dictyosomes, ribosomes, microtubules and plasmodesmata are also distinguishable (**2.14**). Most of these organelles are membrane-bounded (**2.4**, **2.8**, **2.12** to **2.14**, **2.16**).

In this Chapter only the fine-structural features of plant cells are considered, whereas the histological structure of differentiated cells and tissues is discussed in Chapter 3.

Cell membranes

Substances located exterior to the plasmalemma, or in the cytosol surrounding membrane-bounded organelles (**2.4**, **2.14**), cannot mix freely with the materials localised internally because these membranes are semi-permeable. Membranes consist of a lipid bilayer (**2.16**), with the interspersed proteins and complexes forming the molecular pumps, enzymes and other structural components. Some proteins are large and project onto the surface of the membrane (**2.16** to **2.18**). Differing types of organelles normally remain discrete within the cell since their membranes vary somewhat in individual structure.

The plasmalemma (**2.12**, **2.13**, **2.16**) and the membranes of mature dictyosome cisternae and vesicles (**2.8**, **2.19**) are generally the thickest membranes of the cell and measure about 10nm wide. When viewed in transverse section (in chemically fixed material) membranes usually show a tripartite appearance (**2.19**, **2.20**); but in freeze-fractured specimens the plasmalemma (**2.17**, **2.18**) and other membranes (**2.21**, **2.22**) show numerous particles which probably represent protein complexes (**2.16**).

The plasmalemma adjacent to the plant cell wall sometimes reveals hexagonal arrays of particles which are possibly the sites of cellulose microfibrillar synthesis (**2.16**). Likewise in the yeast *Saccharomyces*, chitin microfibrils in the wall apparently link with particles in the plasmalemma (**2.18**).

Nucleus

The genetic material of the cell is primarily located in the nucleus (**2.1**, **2.14**, **2.23**). The non-dividing nucleus is bounded by an envelope composed of the outer and inner membranes (**2.2**, **2.14**, **2.22**). These are separated by a perinuclear space ca. 20 nm wide, but are confluent at the margins of the abundant nucleus pores (**2.14**, **2.21**, **2.22**). These pores are ca. 70 nm wide but are apparently partly occluded by a complex fibrillar-particulate network. In the meristematic cell (**2.4**, **2.23**) the nucleus may occupy a half or more of the volume of the protoplast but this ratio rapidly decreases as the cell increases in size, with the individual small vacuoles (**2.4**) expanding and fusing to form a large central vacuole (**2.24**). The nucleus sometimes becomes highly lobed (**2.15**, **2.23**) and in elongate, narrow cells may be spindle-shaped.

The chromatin (DNA complexed with histones) is not organised into chromosomes in the interphase cell of higher plants but regions of densely-staining heterochromatin and lighter euchromatin are often visible (**2.14**, **2.23**, **2.24**). One to several nucleoli (**2.5**, **2.25**) occur within the nucleoplasm; these contain stores of ribosome precursors (**2.26**) which apparently migrate into the cytoplasm via the nuclear pores. Prominent vacuoles are sometimes evident within the nucleoli (**2.15**, **2.25**, **2.26**). The nuclei of meristematic cells are usually diploid; but DNA replication in interphase is not necessarily followed by nuclear division and in actively metabolising tissues the cells are frequently polyploid.

From the onset of mitosis (prophase) the chromatin is organised into discrete chromosomes which become aligned at metaphase (**2.27**) on the equator of the mitotic spindle. Meanwhile the nucleolus has disappeared and the nuclear envelope fragmented. The fibres of the mitotic spindle, which are just visible at LM level (**2.28**), actually consist of aggregated bundles of microtubules. These are generated at the poles of the spindle and attach to chromosomes and the fibres run between the two poles.

At metaphase each chromosome consists of two chromatids joined at their kinetochores, while microtubules are linked to each kinetochore (**2.27**). At anaphase the sister chromatids separate along the fibres to opposite poles of the mitotic spindle (**2.29**). Finally, at telophase, the chromatin becomes dispersed so that discrete chromosomes are no longer visible. Each nucleus becomes invested by an envelope, the nucleoli reappear whilst a cell plate (new cell wall) separates the two progeny nuclei (**2.25**).

Plastids

A variety of plastids with differing metabolic significance occurs in plants (**2.30**) but, within a particular cell, generally only one form is present (**2.24**). However, in dedifferentiating cells both the mature and juvenile types may occur (**2.15**, **2.32**). All plastids are bounded by a membranous envelope (**2.30**, **2.33**); the outer membrane is continuous but the inner membrane sometimes shows invaginations into the matrix (stroma). The latter contains ribosomes (70s) and circular DNA, which are both chemically distinct from those of the nucleus, while starch and lipid droplets (plastoglobuli) are frequently present (**2.12**, **2.30**, **2.33**). Internal membranes usually occur in the stroma and these sometimes form complex con-

figurations (**2.30**, **2.33**). Plastid interconversions are common (**2.30**), for instance amyloplasts can turn green (**2.34**) and form chloroplasts, while the latter may divide to form young chloroplasts (**2.30**, **2.35**) or senesce and give rise to chromoplasts (**2.36**).

Proplastids

These precursors of other plastids are usually 1–2 µm in maximal width and may be rounded or amoeboid and they contain few internal membranes (**2.37**). Proplastids occur in meristematic cells (**2.2**, **2.37**) and in the root apex up to 40 are present per cell. Proplastids apparently divide and their population remains more or less stable in the cells formed during the numerous divisions within the apical meristems. However, as the derivatives of the apical cells grow and differentiate (**1.18**) their plastid populations generally increase, and the plastids develop into the various types characteristic of different tissues and organs of the mature plant (**2.30**).

Chloroplasts

At maturity these are usually oval to lenticular and 5–10 µm in length (**2.24**, **2.33**, **2.35**). In the leaf mesophyll (**2.24**) up to 50 chloroplasts per cell are commonly present, but in some species they are even more abundant. The membranes of the envelope are separated by a space 10–20 nm wide (**2.33**) and the inner member sometimes shows connections with the thylakoid membranes in the stroma. These membranes are predominantly orientated parallel to the long axis of the plastid (**2.33**, **2.35**) and are normally elaborated into a complex three-dimensional, photosynthetic system. This consists of grana interconnected by stromal lamellae or frets (**2.33**, **2.35**, **2.38**).

Each granum consists of 2–100 flattened and stacked membranous discs and each granal membrane encloses an intrathylakoidal space (**2.33**, **2.39**). The sub-structure of the granal membranes is highly complex, with the chlorophyll molecules being integral components, and their functioning in the light reactions of photosynthesis is under intensive research. The major proteinaceous component in the stroma is the enzyme ribulose diphosphatate carboxylase.

Starch is commonly present in the stroma of the chloroplasts (**2.24**, **2.35**) but represents a temporary store of excess carbohydrate. Plastoglobuli (**2.33**, **2.35**) are frequent and contain pigments concerned in electron transport while phytoferritin (**2.40**) deposits are sometimes observed. During leaf development the chloroplasts increase in number per cell by division (**2.30**, **2.35**).

Etioplasts

In most flowering plants grown in the light the proplastids rapidly differentiate into chloroplasts in the young leaves (**2.35**), but in dark-grown plants etioplasts develop (**2.30**, **2.41**). These contain an elaborate membranous prolamellar body with radiating lamellae but, on exposure to light, these rapidly form a granal-fretwork system and protochlorophyllide is converted to chlorophyll. In some plants, for example grasses, the leaf primordia are tightly ensheathed by the older leaves (**1.23**); in such a darkened internal environment etioplasts initially differentiate in the mesophyll but the leaf blade turns green as it grows out free from the enclosing leaf bases.

Amyloplasts

In storage parenchyma cells long-term deposition of starch occurs in the amyloplasts (**2.31**, **2.32**) in which internal membranes are few, but one to numerous starch grains occur (**2.42**). The starch consists of varying proportions of amylose and amylopectin and is deposited in layers which may be visible at LM level (**2.32**). Amyloplasts in the potato tuber (*Solanum*) reach 20 µm in width and the development of several large starch grains within the plastid often causes the envelope to break and release the grains into the cytoplasm (**2.31**).

When starch is mobilised in the germinating cotyledon of *Phaseolus* (bean), the grain is initially digested from the centre and fragmented grains may be visible in the cytoplasm (**2.43**). In root cap cells, nodal regions and sometimes elsewhere (**2.44**), large sedimented amyloplasts occur and are apparently concerned with gravity perception. Amylochloroplasts (**2.4**) are common in the young shoot; these show thylakoids but also contain prominent deposits of storage starch. In plastids of *Helianthus*, carbohydrate is stored as inulin.

Chromoplasts

Yellow, red and orange plastids are designated as chromoplasts and these accumulate a variety of carotenoid pigments (in globular or crystalline form) which colour many flowers and fruits (**1.6**). The changing foliage colours of deciduous trees prior to leaf fall (**2.45**) are caused by the degeneration of the thylakoid system of the chloroplasts, with the carotenoids accumulating in numerous plastoglobuli (**2.36**). However, many chromoplasts do not represent degenerate chloroplasts but develop directly from proplastids or via amyloplasts (**2.30**, **2.46**).

Leucoplasts

In the epidermis of the green shoot (**2.47**) and the mesophyll of variegated leaves, non-pigmented leucoplasts often occur. These contain few internal membranes and little or no starch.

Mitochondria

Aerobic respiration occurs in these organelles and, as with plastids, mitochondria show nucleoid regions containing circular DNA (**2.48**) and ribosomes. Mitochondria are circular, elliptical or irregular in shape (**2.2**, **2.48**, **2.49**). They are delimited from the cytoplasm by an envelope whose inner membrane shows frequent invaginations into the stroma to form irregular inflated tubules (**2.48**) or cristae (**2.49**). The outer membrane has a high lipid content and the inner membrane contains enzymes of the electron transport chain, while the Krebs cycle enzymes mainly occur within the stroma. Mitochondrial cristae (**2.49**) are especially well developed in rapidly respiring tissues. Although a sectioned cell may show numerous small mitochondria which are one to several micrometres wide (**2.49**), such apparently separate organelles may represent segments of larger polymorphic individuals.

Endoplasmic reticulum (ER)

This system ramifies throughout the protoplast (**2.14**) and in plants it is commonly cisternal (lamellar) in form (**2.48**, **2.50**). It is delimited by a single membrane (**2.48**) which is often confluent with the outer nuclear membrane and is also apparently continuous with the central (desmotubular) component of plasmodesmata (**2.13**, **2.14**). In rough endoplasmic reticulum (RER) the outer surface of the membrane is studded with ribosomes of 17–20 nm diameter (**2.48**) which are somewhat larger than the mitochondrial and chloroplast ribosomes.

The proteins synthesised by the ribosomes sometimes accumulate within the lumen of the associated ER and may be transported in the ER to other sites in the cell. Smooth endoplasmic reticulum (SER) lacks ribosomes and is particularly concerned with lipid synthesis. Extensive tracts occur in oily seeds. The ER system is believed to provide the essential proteinaceous and lipidic components for the other membranous systems of the protoplast.

Golgi apparatus

This comprises several to numerous discrete membranous dictyosomes (golgi bodies) per cell (**2.8**, **2.19**, **2.51**). Each consists of a plate-like stack (**2.14**, **2.52**) of smooth cisternae 1–2 µm in diameter. Their margins are frequently branched into a tubular network (**2.51**) which proliferates a number of small vesicles. The cisternae are separated from each other by about 10 nm, but individual dictyosomes remain intact when isolated from the cell.

In the longitudinal view of a dictyosome a polarity is sometimes evident (**2.52**). At its forming face the cisternae are thought to be reconstituted from vesicles budded off from adjacent membranes of the ER (**2.14**, **2.52**). The cisternal membranes progressively increase in thickness across the dictyosome and at its maturing face the cisternae frequently become concave, with vesicles budding off from their margins (**2.14**, **2.51**, **2.52**).

The vesicles apparently migrate and fuse with the plasmalemma where the contents are voided into the apoplast. Within the dictyosome, proteins derived from the RER combine with various sugars; the secreted vesicles contain carbohydrates and glycoproteins concerned with cell wall synthesis (**2.8**), mucilage (**2.19**), or nectar secretion.

Dictyosomes are particularly abundant in cells actively undergoing extensive wall thickening or forming new dividing walls (**2.8**, **2.53**); in root cap cells, where cisternae often become inflated with mucilage and sloughed off whole (**2.19**); and in glandular cells. In actively secreting cells a vast potential excess of membranous components (derived from the dictyosomes) arrives at the plasmalemma, but some of this material is apparently returned to the cytoplasm in the 'coated' vesicles (**2.14**).

Vacuole

In the meristematic cell a number of small vacuoles occur, each bounded by the membranous tonoplast (**2.2**, **2.4**, **2.13**). However, during cell growth these vacuoles massively enlarge and fuse so that up to 90% of the volume in a parenchyma cell is occupied by the vacuole (**1.18**), while the cytoplasm and other organelles are mainly peripheral (**2.24**, **2.42**). The vacuole contains various solutes (normally at about 0.5 M concentration) and its consequent turgidity greatly contributes to the turgor of the whole protoplast.

The vacuole also contains a number of hydrolytic enzymes and the tonoplast breaks down during differentiation of sclerenchyma and tracheary cells (**2.7** to **2.9**). The enzymes which are liberated digest the protoplasts so that only the walls remain intact (**1.4**, **1.11**). Vacuoles may contain anthocyanins and other pigments and also become modified as protein bodies in a number of seeds (**2.43**, **2.54**).

Microbodies

These small, membrane-bounded bodies are of two types. Peroxysomes occur in close proximity to chloroplasts (**2.38**, **2.39**); they contain a variety of enzymes which oxidise the glycolic acid resulting from photorespiration and then return glyoxylic acid to the chloroplasts. Glyoxysomes occur in fatty seeds and contain enzymes catalysing fatty acid breakdown to release energy during germination, while the hydrogen peroxide produced is broken down by peroxidase.

Ribosomes

The cytoplasmic ribosomes are 17–20 nm wide and occur both floating free within the cytosol (cytoplasmic ground substance) as well as attached to the outer surfaces of the RER (**2.48**, **2.50** and **2.51**). Ribosomes contain RNA and protein and these are composed of two sub-units which are synthesised in the nucleolus (**2.26**) but subsequently combine in the cytoplasm. Ribosomes are especially dense in cells which are rapidly synthesising protein where they frequently occur in clusters termed polysomes (**2.48**). The cytoplasmic ribosomes (80S) are slightly larger and biochemically distinct from those located in the plastids and mitochondria (70S).

Microtubules and microfilaments

The microtubules are proteinaceous structures about 25 nm wide, with a hollow core 12 nm wide, and may be up to several micrometers long (**2.55**). In the non-dividing cell they are normally located adjacent to the plasmalemma (**2.2**, **2.55** and **2.56**), but at nuclear division these become reassembled and aggregate into the fibres of the mitotic spindle (**2.28**, **2.56**). When mitosis is completed, the microtubules apparently guide dictyosome vesicles to the equatorial region of the spindle where the vesicles fuse to form the cell plate (**2.53**, **2.57**, **2.58**). It has been suggested that the peripheral microtubules are concerned with the orientation of the cellulose microfibrils which are being formed in the young wall on the outside of the plasmalemma (**2.16**, **2.17**); however, the evidence for this is equivocal.

Another smaller type of proteinaceous element has sometimes been observed in the plant cell. This is termed a microfilament and is about 7 nm in width. In the green alga *Nitella*, the microfilaments are concerned with cytoplasmic streaming; in pollen tubes of flowering plants they apparently guide vesicles concerned with wall synthesis through the cytosol to the growing tip of the tube.

The cell wall

Plant protoplasts are normally enclosed by a wall (2.3, 2.4) which gives rigidity and protection to the cell but, unless impregnated with fatty materials, does not prevent water and solutes diffusing across it to the plasmalemma. The walls of adjacent cells are cemented together by a common middle lamella (2.11, 2.12) so that plant cells are immobile, although fibres and some other cells elongate by tip growth and intrude between the neighbouring cells.

Following cell division the progeny usually undergo vacuolation growth and their primary walls also expand. Commonly in parenchymatous tissue the middle lamellae break down at the sites where several cells connect to each other and intercellular spaces develop (2.59). When expansion growth ceases some cell types undergo deposition of a secondary wall (2.7 to 2.10). The constitution of the secondary wall components and the orientation of its cellulose microfibrils (2.60) is markedly different from that of the primary wall.

Primary wall

Cytokinesis normally immediately follows mitosis and so the two progeny are divided by a common cell plate (2.25). The plate first appears at the equator of the mitotic spindle (2.56) and then advances centrifugally (2.56 to 2.58, 2.61) to fuse with the mother cell wall. The plate is formed from fusing dictyosome vesicles (2.53) and is delimited by a plasmalemma derived from the dictyosome membranes (2.57, 2.58). Strands of ER penetrate the plate and these form the central desmotubular component of the plasmodesmata (2.13, 2.14) which connect the protoplasts of adjacent cells. The unthickened cell plate constitutes the middle lamella common to both daughter cells (2.57, 2.58). It is largely composed of pectic substances and subsequently a thin primary cell wall is deposited on both surfaces of the plate (2.12, 2.25).

When a highly vacuolated cell divides, the growing margins of the cell plate are marked by dense cytoplasm (the phragmosome, 2.61) in which clusters of short microtubules occur (2.56 to 2.58). In some tissues mitosis is not immediately followed by cell plate development: in the endosperm of many species the initially coenocytic cytoplasm (2.5) later becomes divided by freely-forming walls. These often develop in a tortuous pattern and a similar phenomenon occurs in callus tissue (2.62). In transfer cells tortuous invaginations of the primary wall into the protoplast occur (2.63) and such modifications of parenchyma cells are common adjacent to vascular elements.

In thinner areas of the wall plasmodesmata are often clustered together (2.2), with up to 60 present per square micrometer of wall surface, to form pit fields (2.64). When secondary wall deposition occurs, these regions often remain unthickened and give rise to pits closed internally by a pit membrane (primary wall, 2.10, 2.54, 2.65). Where lignification of the wall occurs, it largely restricts the passage of water and nutrients between protoplasts to the non-lignified pitted regions (2.10).

In the tracheary elements of the protoxylem (2.9) a lignified secondary wall is deposited internal to the primary walls in discrete rings or a spiral, but extensive tracts of non-lignified primary wall lie between the thickenings. These non-lignified regions are attacked by hydrolytic enzymes released from the vacuole of the degenerating protoplast, so that frequently only the cellulosic skeleton ('holey' wall) remains in the mature element to indicate the original position of the primary wall (2.66).

The somewhat thickened walls of sieve tubes (2.6), and the thick walls of many storage parenchyma cells (2.54) are primary and do not normally undergo lignification. During differentiation of the sieve elements their protoplasts largely degenerate (2.67), leaving intact the plasmalemma together with modified plastids (2.68, 2.69), mitochondria, endoplasmic reticulum and deposits of P-protein (2.67, 2.68). The end walls of the sieve elements become modified as the amorphous polysaccharide, callose, is deposited within the wall around the plasmodesmata (2.6, 2.67). The desmotubular component of the plasmodesma disappears and eventually a wide pore develops, ranging from 1–15 μm in diameter, but this usually appears plugged in sectional material (2.6, 2.68).

The primary wall contains up to 80% of its fresh weight as water, while the other components are predominantly polysaccharide. Biochemical analysis shows that cellulose constitutes 25–30% by weight of the dried wall while hemicelluloses constitute a further 15–25%, pectic substances up to 35%, and glycoproteins 5–10%.

Cellulose is a polymer of glucose with units linked into long unbranched chains of up to 15,000 monomers. These are laterally hydrogen-bonded to form microfibrils (2.17, 2.70) several micrometres long and 3.0–8.5 nm wide. The microfibrils possess high tensile strength and reinforce the wall in a form analogous to steel rods in reinforced concrete. The other polysaccharide components are non-fibrillar. It is considered likely that the hemicelluloses are linked to the microfibrils by hydrogen bonding. In turn the hemicellulose is covalently attached to a neutral pectin which is linked to an acidic pectin. These non-cellulosic polysaccharides interact with water to form a gel which determines the plasticity of the wall.

In the primary wall newly formed microfibrils may be aligned parallel to each other (2.17), but randomly orientated microfibrils are also common (2.60). Wall matrix substances are synthesised in the ER and dictyosomes and transported via dictyosome vesicles to the newly-forming wall. It seems, however, that the cellulose microfibrils form *de novo* at the plasmalemma (2.16). In yeast the chitin microfibrils of the wall apparently originate from rosettes (aggregates of particles) in the plasmalemma (2.18).

A similar mechanism has been suggested for the formation of cellulose microfibrils in higher plants in which enzyme rosettes, composed of cellulose synthase molecules (2.16), receive activated glucose from the protoplast. The microfibril extends outwards into the new wall surface as more cellulose is synthesised at the rosette. However, in many plants investigated by electron microscopy such rosettes have not been observed on the plasmalemma.

Secondary wall

These walls (2.7, 2.8, 2.60) are typically much less hydrated than the primary wall, and a higher proportion of the polysaccharide content is cellulose. Normally in a mature fibre the microfibrils in the oldest layer of the secondary wall (S1) and the youngest layer (S3) show microfibrils orientated more-or-less transversely to the fibre's long axis (2.60). In the thicker middle layer (S2) they are aligned nearly parallel to this axis. Most fibres are lignified, but in flax (*Linum*, 2.7) this rarely occurs and the multilamellate walls contain about 90% cellulose dry weight.

Mature sclerenchyma and tracheary elements are normally heavily lignified (2.9, 2.10, 2.66). Between 15–35% of their dry weight is composed of lignin which is of high molecular weight and contains various aromatic alcohols. Lignin is chemically inert and provides a resistant, water-proofing barrier around the cellulose microfibrils. The cross-linking of lignin to cellulose provides a strong and rigid cell wall and isolated strands of fibres commonly have a tensile strength comparable to that of a steel wire of the same diameter.

2.1 LS of *Phaseolus vulgaris* (bean) root just behind the apex. This section shows longitudinal files of incipient cortical cells which are still actively dividing. Note the densely-staining interphase nuclei (1) whilst a number of other nuclei are in various stages of mitosis (2). (LM x 375.)

1	Interphase nuclei
2	Nuclei in mitosis

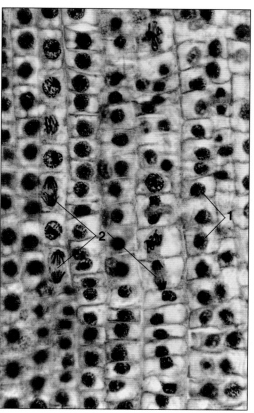

2.1

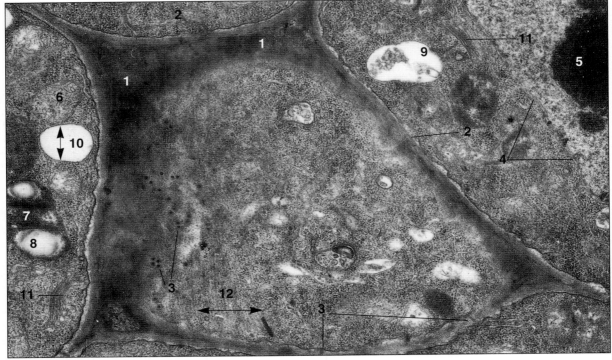

2.2 Meristematic cells from a bud of *Glechoma hederacea* (ground ivy). The protoplast is separated from the external cell wall (1) by the membranous plasmalemma (2), while plasmodesmata (3) penetrate the wall to connect adjacent protoplasts. The nucleus is invested by the two membranes of the nuclear envelope (4) and contains a densely-staining nucleolus (5).

The mitochondria (6) and proplastids (7) are also enclosed by membranous envelopes; these latter organelles are difficult to distinguish from each other in meristematic cells but starch (8) is diagnostic of a plastid. The small vacuoles (9) are delimited from the dense cytoplasm by a single tonoplast membrane (10). Several membranous dictyosomes (11) occur but little endoplasmic reticulum can be distinguished. A number of proteinaceous microtubules (12) are present. (G-Os, TEM x 6,000.)

1	Cell wall
2	Plasmalemma
3	Plasmodesmata
4	Nuclear envelope
5	Nucleolus
6	Mitochondrion
7	Proplastid
8	Starch
9	Vacuole
10	Tonoplast
11	Dictyosome
12	Proteinaceous microtublules

2.3 TS of a procambial strand in a bud of *Glechoma hederacea* (ground ivy). The narrow procambial cells are thin-walled, densely staining and contain relatively large nuclei. These meristematic cells contrast with both the thick-walled, apparently empty, protophloem sieve tubes (1) and the mature protoxylem tracheary elements (2) which show irregularly-thickened secondary walls. Cortex (3). (G-Os, TEM x 1,800.)

1	Protophloem sieve tube
2	Protoxylem tracheary element
3	Cortex

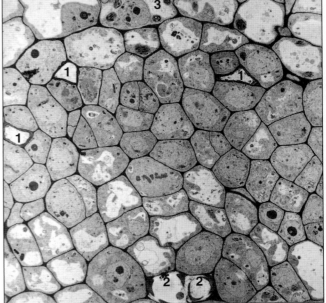

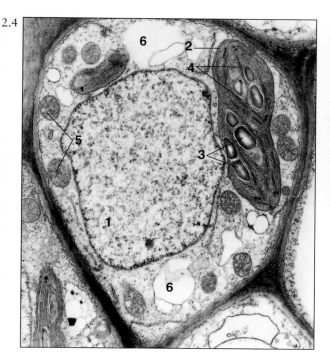

2.4 Bundle sheath cell from the leaf of *Phaseolus vulgaris* (bean). Note: the large nucleus (1), prominent amylo-chloroplast (2) with numerous starch grains (3) and a well developed thylakoid system (4), mitochondria (5), and several small vacuoles (6). (G-Os, TEM x 10,000.)

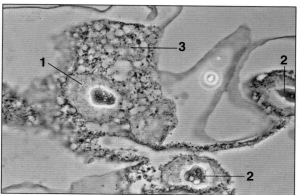

2.5 Coenocytic endosperm surrounding the young embryo of *Phaseolus vulgaris* (bean). Several nuclei (1) with prominent nucleoli (2) lie embedded in vesiculate cytoplasm (3); but no cell walls are present. (G-Os, Phase LM x 355.)

1	Nucleus
2	Nucleolus
3	Vesiculate cytoplasm

1	Nucleus	4	Thylakoid system
2	Amylochloroplast	5	Mitochondrion
3	Starch grains	6	Vacuole

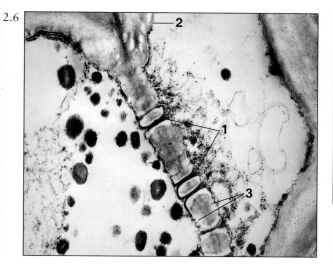

2.6 LS through a sieve plate of *Sorbus aucuparia* (rowan tree) showing narrow pores (1). Although these are apparently blocked, they are thought to be open in life. The protoplasts of the sieve elements are enucleate and the other organelles are degraded, but the plasmalemma (2) is still intact. Callose (3). (G-Os, TEM x 15,500.)

| 1 | Pore | 3 | Callose |
| 2 | Plasmalemma | | |

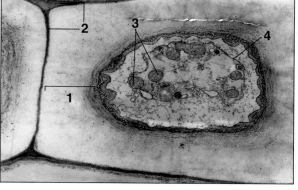

2.7 TS of a phloem fibre from the young hypocotyl of *Linum usitatissimum* (flax). This region of the fibre is nearing maturity and shows a massively thickened, but non-lignified, secondary wall (1) inside the thin primary wall (2). Within the degenerating protoplast a number of mitochondria (3) are distinguishable but the plasmalemma (4) has become detached from the innermost layer of secondary wall. (G-Os, TEM x 1,800.)

| 1 | Secondary wall | 3 | Mitochondrion |
| 2 | Primary wall | 4 | Plasmalemma |

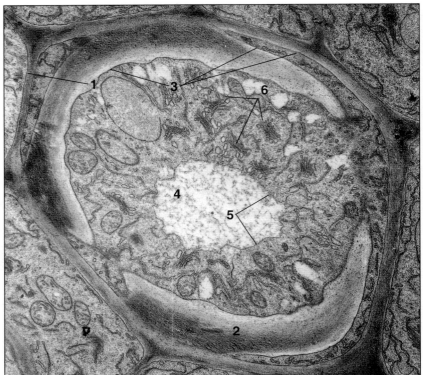

2.8 TS of a helically-thickened, differentiating protoxylem element of *Glechoma hederacea* (ground ivy). The thin primary wall (1) and the thickened, lignified secondary wall (2) are delimited from the protoplast by the plasmalemma (3). The protoplast contains a central vacuole (4) with its tonoplast (5) still intact. Numerous dictyosomes (6) are secreting vesicles into cytoplasm and these are presumed to transport non-cellulosic polysaccharides to the growing wall. Numerous endoplasmic reticulum cisternae are evident but with this fixation the ribosomes are not preserved. (Km, TEM x 9,800.)

1	Primary wall
2	Secondary wall
3	Plasmalemma
4	Vacuole
5	Tonoplast
6	Dictyosomes

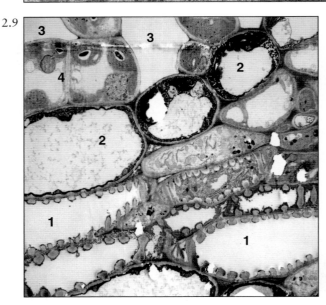

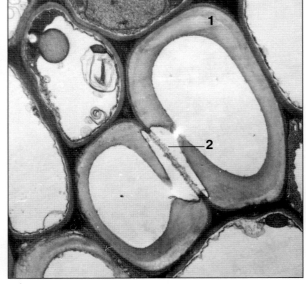

2.9 LS of a veinlet from a leaf of *Sorbus aucuparia* (rowan tree). Several annular protoxylem elements are already mature (1) and without protoplasts. Note the bundle sheath (2) parenchyma cells which separate the xylem elements from the air spaces (3) of the surrounding mesophyll tissue (4). (G-Os, TEM x 2,000.)

1	Annular protoxylem
2	Bundle sheath
3	Air spaces
4	Mesophyll tissue

2.10 TS of xylem from the leaf midrib of *Sorbus aucuparia* (rowan tree). Note the several mature tracheary elements with thick secondary walls (1) and lumina devoid of protoplasts. A prominent bordered pit is visible and the primary wall (2), which forms the pit membrane, is less dense when it is not overlain by secondary wall. (G-Os, TEM x 1,800.)

1	Secondary wall
2	Pit membrane

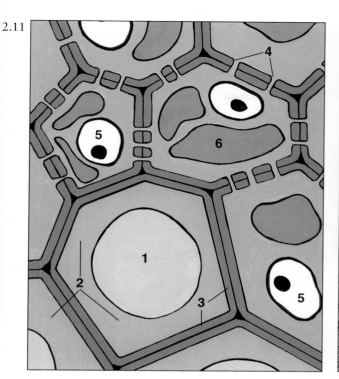

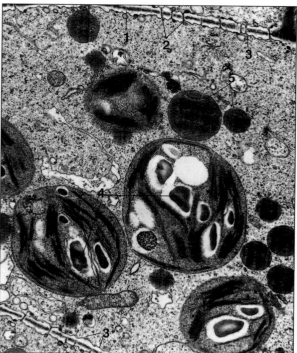

2.11 Diagram to illustrate the symplast and apoplast. Each dead sclerenchyma fibre has an empty central lumen (1) which is enclosed by a thick secondary wall (2) lying within a thin primary wall (3). The living parenchyma cells possess primary walls only (3) and their protoplasts are interconnected by numerous plasmodesmatal channels (4). The protoplasts collectively constitute the symplast while the cell walls and dead cell lumens (plus intercellular space when present) form the apoplast. Nucleus (5), vacuole (6).

2.12 Fine-structure of a mesophyll cell from the leaf of the moss *Polytrichum commune*. The cells are bounded by unthickened primary walls separated by a densely-staining middle lamella (1); note the numerous plasmodesmata (2) traversing the walls and the plasmalemma (3) delimiting the wall from the protoplast. In the cytoplasm large chloroplasts (with starch grains, 4) and lipid vesicles (5) are especially prominent. (G-Os, TEM x 10,500.)

1	Central lumen	4	Plasmodesmatal channels
2	Secondary wall	5	Nucleus
3	Primary walls	6	Vacuole

1	Middle lamella	4	Starch grain
2	Plasmodesmata	5	Lipid vesicles
3	Plasmalemma		

2.13 Root tip cells from *Allium cepa* (onion). Numerous plasmodesmata (1) cross the thin, unstained primary wall. The single membranes of the plasmalemma (2), tonoplast (3), endoplasmic reticulum (4) and dictyosomes (5) are well defined while the double membranes investing the mito-chondria (6) and proplastids (7) can also be discerned. However, neither ribosomes nor microtubules are preserved with this fixative. (Km, TEM x 20,500)

1	Plasmodesmata
2	Plasmalemma
3	Tonoplast
4	Endoplasmic reticulum
5	Dictyosomes
6	Mitochondria
7	Proplastid

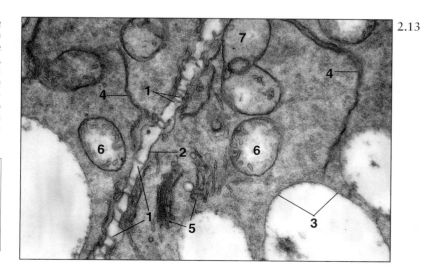

2.14

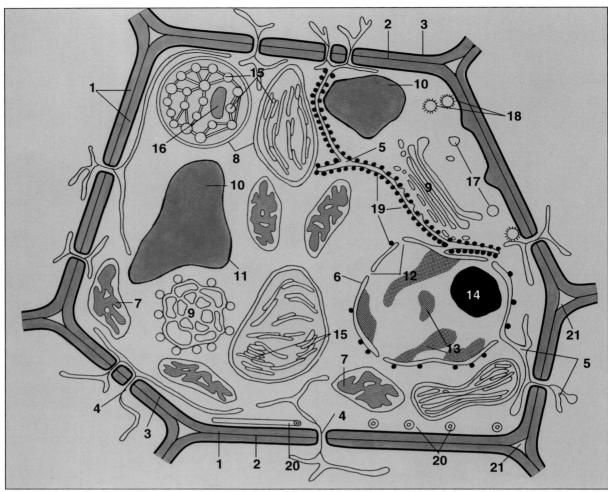

2.14 Diagram of the fine-structure of a relatively undifferentiated plant cell. The protoplast is bounded by a thin primary wall (1) with a median middle lamella (2). The plasmalemma (3) encloses the protoplast and this membrane also lines the plasmodesmatal pores. Narrow membranous desmotubules (4) traverse these pores and link the endoplasmic reticulum (5) of adjacent protoplasts. Although only partially indicated, the outer surfaces of the endoplasmic reticulum and outer nuclear membrane are normally covered by ribosomes; the numerous free cytoplasmic ribosomes are not shown in this diagram.

The majority of cell organelles are membrane-bounded; two membranes enclose the nucleus (6), mitochondria (7) and chloroplasts (8), while the cisternae of the endoplasmic reticulum (5) and dictyosomes (9) are delimited by single membranes. The vacuoles (10) are also bounded by a single tonoplast (11). The nuclear envelope is linked to the endoplasmic reticulum, while at the numerous nuclear pores (12) the inner and outer membranes are confluent. Within the interphase nucleus denser DNA-rich areas of heterochromatin (13) occur and a large nucleolus (14) is evident. Large chloroplasts are present showing well-developed photosynthetic granae (15) and starch (16). The inner membrane of the mitochondrial envelopes shows convoluted tubular, or sometimes plate-like, invaginations.

The dictyosomes (illustrated in both longitudinal and transverse views) show a polarity with the maturing face budding off numerous vesicles (17) which apparently migrate through cytoplasm to fuse with the plasmalemma (arrows). The coated vesicles (18) probably return surplus membranous material to be recycled by the protoplast. The principal non-membranous organelles within the cell are the ribosomes (19) and the proteinaceous microtubules (20). At interphase the latter lie adjacent to the plasmalemma. Note that the middle lamella lying at the angles of the cell wall is beginning to break down to form intercellular spaces (21).

1	Primary wall
2	Middle lamella
3	Plasmalemma
4	Desmotubules
5	Endoplastic reticulum
6	Nucleus
7	Mitochondria
8	Chloroplasts
9	Dictyosomes
10	Vacuoles
11	Tonoplast
12	Nuclear pores
13	Heterochromatin
14	Nucleolus
15	Granae
16	Starch
17	Dictyosome vesicles
18	Coated vesicles
19	Ribosomes
20	Proteinaceous microtubules
21	Intercellular spaces

2.15 Polymorphic nucleus from a de-differentiating cell of *Phaseolus vulgaris* (bean). The nucleus contains several prominent vacuoles (1) and is surrounded by a dense cluster of small amyloplasts (2). (G-Os, LM x 1,200.)

1	Nuclear vacuoles
2	Amyloplasts

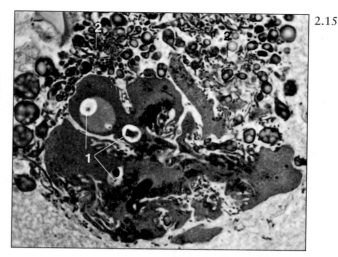

2.16 Model of a generalised plant membrane (**A**) and plasmalemma (**B**). A membrane is thought to consist of a bilayer of phospholipids with their hydrophilic heads (1) outermost, in which proteins are interspersed. Some proteins are confined to the membrane surfaces (2) while others (3) traverse the bilayer.

At the plasmalemma (**B**) rosette protein complexes (4, composed of six cellulose synthase molecules), also span the membrane. Here cellulose precursor molecules are taken up from the cytoplasm, while cellulose microfibrils are extruded into the cell wall on the outer face of the plasmalemma.

1	Phospholipid heads	4	Cellulose synthase
2	Surface proteins		molecules
3	Traversing proteins		

2.17 Interface between plasmalemma and cell wall in the root tip of *Lycopersicon esculentum* (tomato). The plasmalemma shows numerous small particles (presumably protein molecules) but no rosettes are apparent. Much of the water in the primary wall has sublimed away to reveal several layers of wall in which numerous, predominantly parallel, cellulose microfibrils occur. The specimen should be viewed from its direction of shadowing (wide arrow). (F-E, TEM x 30,000.)

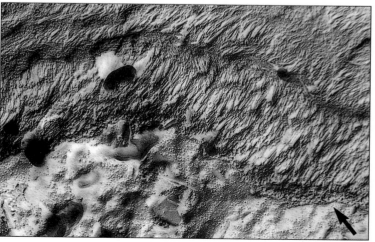

2.18   The plasmalemma of *Saccharomyces* (baker's yeast). Its surface shows numerous particles, many of which are tightly grouped into rosettes (1). A portion of the chitinaceous cell wall (2) is visible and in some regions (3) wall microfibrils appear to originate from the rosette particles. Note also the variously orientated grooves in the plasmalemma: the specimen should be viewed from the direction of its shadowing (wide arrow). (F-E, TEM x 61,000.)

1	Rosettes	3	Wall microfibrils
2	Chitinaceous cell wall		

2.19   Detail from a root cap cell of *Zea mays* (maize) showing a dictyosome (1) with hypertrophied cisternae. These cisternae are bounded by a tripartite membrane and contain a finely granular dense material (2, probably mucilage) which is excreted into the apoplast after the cisternae fuse with the plasmalemma. (G-Os, TEM x 24,000.)

1	Dictyosome
2	Granular material

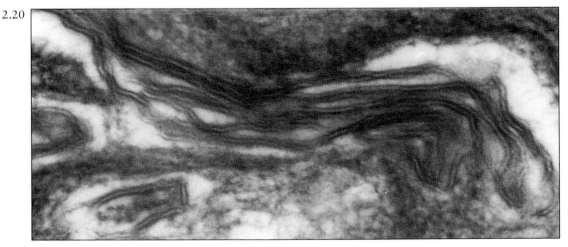

2.20   Membrane complex associated with the plasmalemma of *Andrographis paniculata*. Note the tripartite appearance of transversely-sectioned membranes which show a central translucent layer between two dense, but narrower, outer layers. The surrounding matrix represents the cell wall. (G-Os, TEM x 205,000.)

2.21 Surface view of a nucleus from the root tip of *Lycopersicon esculentum* (tomato). Note the numerous nuclear pores (1) connecting the nucleoplasm and cytoplasm. The specimen should be viewed from the direction of shadowing (wide arrow). Endoplasmic reticulum (2), vacuole (3). (F-E, TEM x 14,000.)

1	Nulear pores
2	Endoplasmic reticulum
3	Vacuole

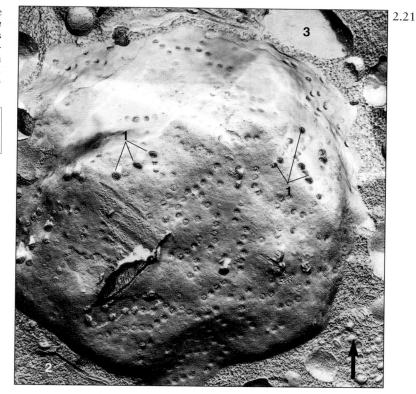

2.21

2.22 Detail of a nucleus from the root tip of *Lycopersicon esculentum* (tomato). The inner (1) and outer nuclear membrane (2) both show numerous pores (3) but the small particles (probably proteins) are much scarcer in the inner membrane. The specimen should be viewed from the direction of shadowing (wide arrow). (F-E, TEM x 33,500.)

1	Inner nuclear membrane
2	Outer nuclear membrane
3	Pores

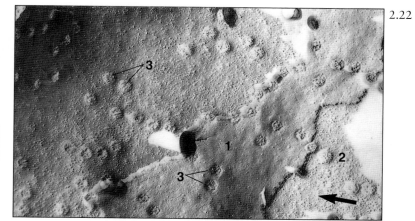

2.22

2.23 Large polymorphic nuclei from callused root tissue of *Pisum sativum* (pea). Note the dense areas of heterochromatin (1) visible internally. Plastids (2), cell wall (3). (G-Os, TEM. x 3,000.)

1	Heterochromatin
2	Plastid
3	Cell wall

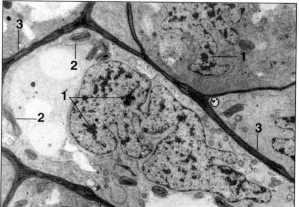

2.23

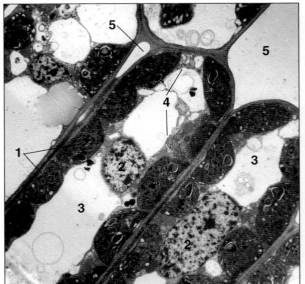

2.24

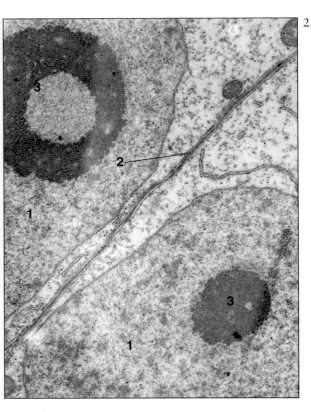

2.2?

2.24 LS of the palisade mesophyll in the lamina of *Sorbus aucuparia* (rowan tree). Each thin-walled palisade cell shows a dense cytoplasmic layer packed with chloroplasts (1) and a single nucleus (2), while the extensive central vacuole (3) is delimited from the cytoplasm by the tonoplast (4). Note the dense heterochromatin in the nuclei and the prominent intercellular spaces (5). (G-Os, TEM x 2,400.)

1	Chloroplasts	4	Tonoplast
2	Nucleus	5	Intercellular spaces
3	Central vacuole		

2.25 Dividing cells of the root of *Pisum sativum* (pea). Note the thin, newly-formed cell wall between the two progeny nuclei (1) resulting from mitosis. The densely-stained middle lamella (2) separates the translucent primary walls of the daughter cells to either side. Nucleolus (3). (G-Os, TEM x 8,500.)

1	Nuclei	3	Nucleolus
2	Middle lamella		

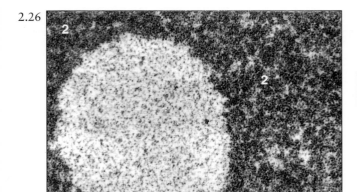

2.26

2.26 Detail of a nucleolus from the root of *Pisum sativum* (pea). Note its large vacuole (1) whose empty appearance contrasts with the crowded particles (2) of the surrounding nucleolus. (G-Os, TEM. x 28,000.)

1	Vacuole
2	Crowded particles

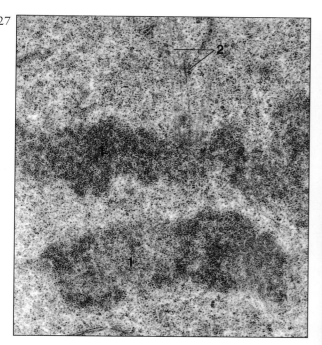

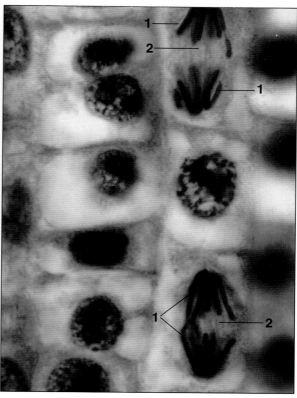

2.27 Detail of a dividing nucleus from *Pisum sativum* (pea) root. The mitotic spindle is sectioned longitudinally at metaphase and shows a pair of densely-staining chromosomes (1) with attached kinetochore microtubules (2). (G-Os, TEM x 21,500.)

1	Chromosomes
2	Kinetochore microtubules

2.28 LS of *Phaseolus vulgaris* (bean) root just behind the apex. Note the two cells in which the nuclei are at the telophase stage of division with densely-staining chromosomes (1) at either pole of the mitotic spindle. The spindle fibres (2) represent microtubules which are grouped in bundles and therefore are visable at light microscopic level. (LM x 1,300.)

1	Chromosomes
2	Spindle fibres

2.29 LS through a mitotic nucleus from *Pisum sativum* (pea) root. At anaphase the densely-staining chromosomes (1) have separated leaving the equator region (2) devoid of large organelles. The nuclear envelope broke down at prophase, but numerous strands of endoplasmic reticulum (3) lie in the cytoplasm together with other organelles. (G-Os, TEM x 7,000.)

1	Chromosomes
2	Equator region
3	Endoplasmic reticulum

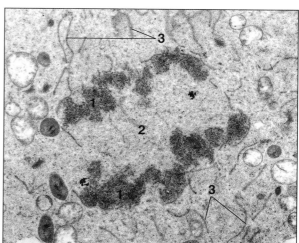

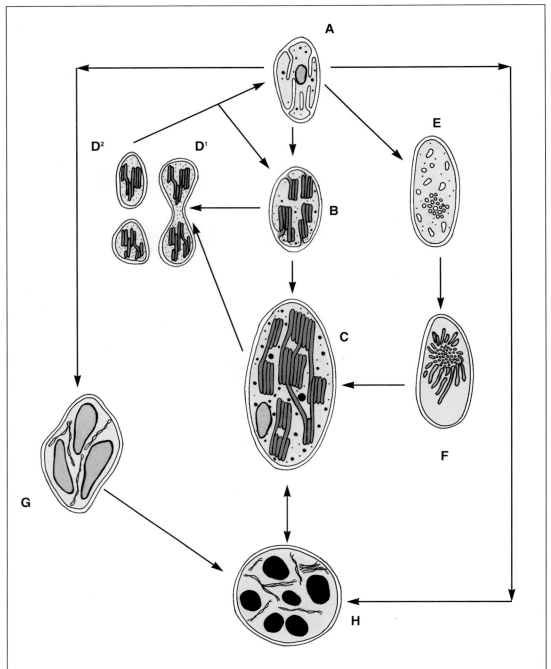

2.30 A to **H**.   Diagram showing the principal pathways of plastid ontogeny. In young green shoots the typical mode is from non-pigmented proplastids (**A**) to mature green chloroplasts (**C**) via young chloroplasts (**B**). Plastid replication may occur at any stage but is particularly common in proplastids and young chloroplasts (**D¹–D²**). In dark-grown shoots (or in cells shielded from light by overlying tissue) etioplasts (**E**) form but on exposure to light these rapidly differentiate (**F**) into chloroplasts (**C**).

In roots (and in most epidermal cells of the shoot) non-pigmented leucoplasts develop from proplastids and amyloplasts (**G**). Chromoplasts (**H**) contain red, orange or yellow pigments; these plastids either develop directly from proplastids or form from amyloplasts or degenerate chloroplasts. Both mature chloroplasts and amyloplasts may be induced to divide (and their derivatives sometimes revert to proplastids) by wounding the tissue in which they occur. The transformation of chromoplasts into chloroplasts has also been reported.

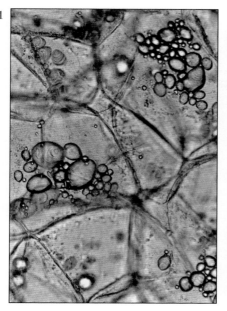

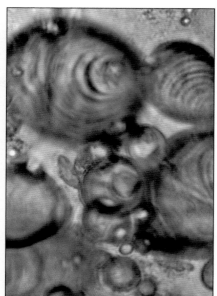

2.31, 2.32 Storage tissue of the greening tuber of *Solanum tuberosum* (potato). **2.31** shows several highly vacuolated parenchyma cells containing amyloplasts with very large starch grains. **2.32** illustrates a group of amyloplasts in greater detail showing the concentric lamellae in the large starch grains; note also the small, pale-green amylochloroplasts. (Fresh section, LM, **2.31** x 145, **2.32** x 575.)

2.33 Young chloroplast from *Crambe maritima* (seakale) sectioned through its long axis. The two membranes of the chloroplast envelope (1) are distinct, while internally flattened membranous thylakoids are stacked into grana (2) interconnected by frets. A number of peripheral membranous vesicles (3) and a densely staining group of plastoglobuli (4) are also visible. (G-Os, TEM x 77,000.)

1	Chloroplast envelope
2	Grana
3	Membranous vesicles
4	Plastoglobuli

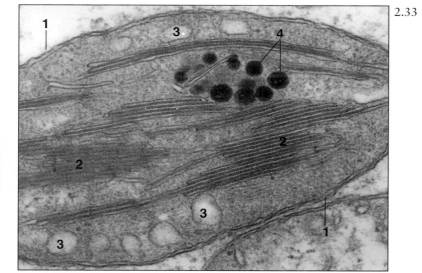

2.34 Storage tissue of the greening tuber of *Solanum tuberosum* (potato). Note the numerous green chloroplasts which have developed after prolonged exposure to light. (Fresh section, LM x 450.)

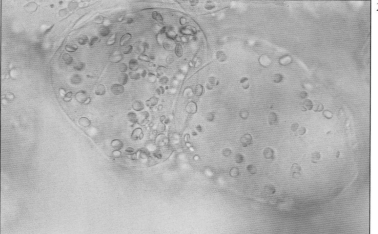

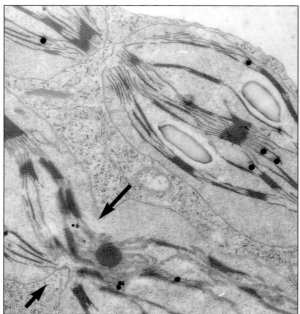

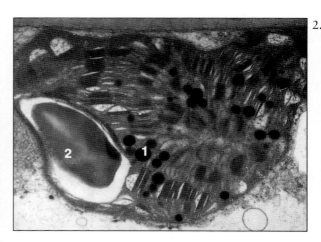

2.36 Chloroplast undergoing transformation into a chromoplast in the leaf of *Sorbus aucuparia* (rowan tree). The leaf was sampled in autumn; note the degenerating grana and the numerous plastoglobuli (1) in which carotenoid pigments accumulate to give the yellow-, orange- and red-coloured autumn foliage of deciduous species. Starch grain (2). (G-Os, TEM x 14,500.)

2.35 Dividing chloroplasts from mesophyll tissue in *Linum usitatssimum* (flax). One plastid has a median isthmus (arrows) which indicates the prospective plane of division, while the other two plastids have just separated. (G-Os, TEM x 7,500.)

1	Plastoglobuli
2	Starch grain

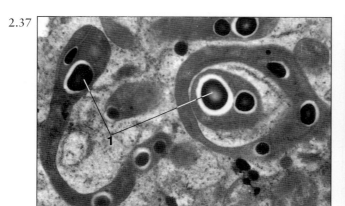

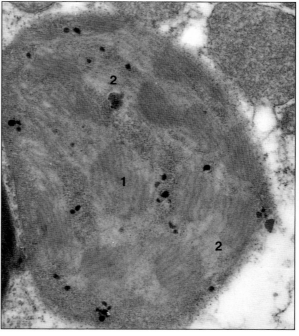

2.37 Proplastids from de-differentiating cotyledonary tissue of *Phaseolus vulgaris* (bean). Few membranes are evident in the stroma but the numerous small starch grains (1) clearly define these as plastids. (G-Os, TEM x 7,000.)

1	Starch grains

1	Granal membranes
2	Fret system

2.38 Small chloroplast from *Crambe maritima* (seakale) sectioned across its shorter axis. Note the discoidal shape of the granal membranes (1) and the tubular fret system (2) which interconnects the grana. (G-Os, TEM x 18,000.)

2.39 Longitudinally-sectioned chloroplast from *Linum usitatissimum* (flax). Note the well developed granal membranes (1) within the stroma and the two membranes of the investing envelope (2). Closely associated with the plastid is a large peroxysome, with a granular matrix (3), which is bounded by a single membrane (4). Tonoplast (5), plasmalemma (6). (G-Os, TEM x 29,000.)

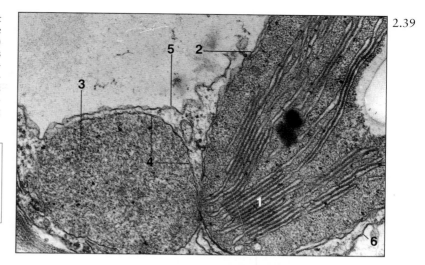

2.39

1	Granal membranes
2	Chloroplast envelope
3	Peroxysome
4	Single membrane of peroxysome
5	Tonoplast
6	Plasmalemma

2.40 Plastid of *Crambe maritima* (seakale) showing crystalline phytoferretin within the stroma. The two membranes constituting the plastid envelope (1) are clearly evident. (G-Os, TEM x 111,500.)

1	Plastid envelope

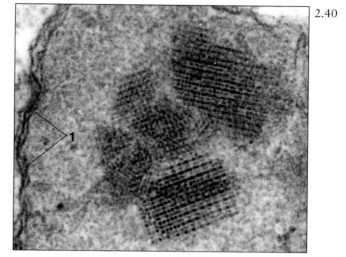

2.40

2.41 Greening etioplast in the stem of *Glechoma hederacea* (ground ivy). From the vestiges of the prolamellar body (1) a granal/fret membranous system (2) is developing. Starch (3), plastid envelope (4). (G-Os, TEM x 47,500.)

1	Prolamellar body
2	Granal/fret membranous system
3	Starch
4	Plastid envelope

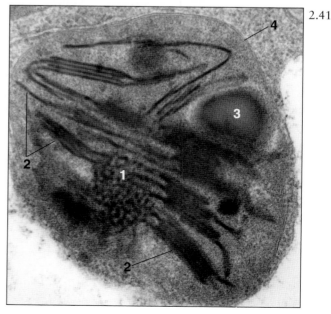

2.41

45

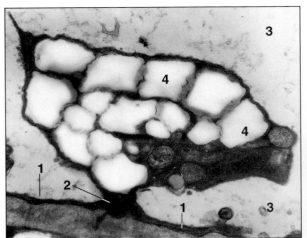

2.42

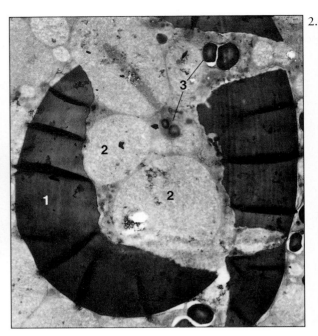

2.4

2.42 Large amyloplast from callus tissue of *Andrographis paniculata*. The plastid is connected to the cytoplasm lining the cell wall (1) by an isthmus of cytoplasm (2) which also invests the amyloplast and separates it from the large vacuole (3). Within the plastid few membranes are evident but numerous starch grains (4) occur. (G-Os, TEM x 12,000.)

1	Cytoplasm lining cell wall
2	Connecting cytoplasm
3	Vacuole
4	Starch grains

2.43 Starch grain fragments from the germinating cotyledon of *Phaseolus vulgaris* (bean). The densely staining starch grain has apparently been digested from the centre leading to its break-up into several pieces (1). Note also the membrane-bounded protein bodies (2), now largely empty, and the small amyloplasts (3). (G-Os, TEM x 3,500.)

1	Starch grain fragments	3	Amyloplasts
2	Protein bodies		

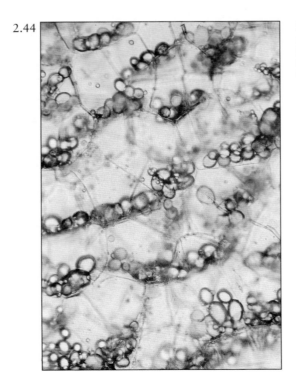

2.44

2.44 LS through an excised segment of *Solanum tuberosum* (potato) tuber. This was incubated in a vertical position for four weeks; note that the amyloplasts have sedimented and lie against the lower walls of the parenchyma cells. (Fresh section, LM x 100.)

2.45 Fallen autumn leaves which have drifted into a bay of the River Kelvin. Note the range of leaf colours, with a few which are still green, through to orange and yellow; the latter colours result from the accumulation of carotenoids within the degenerate chloroplasts.

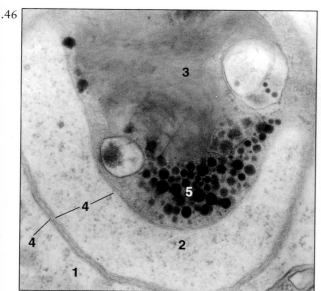

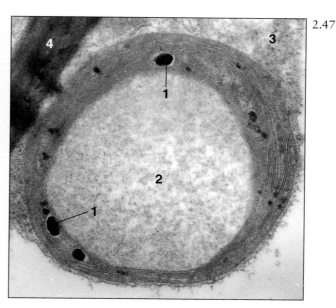

2.46 Chromoplast from the root of *Daucas carota* (carrot). The plastid is invested by an envelope but an extensive intrusion from the surrounding cytoplasm (1) forms a prominent enclave (2). The plastid stroma (3) is reduced here to a very narrow layer so that the plastid envelope (4) is apparently composed of four membranes. Plastoglobuli (5). (G-Os, TEM x 32,000.)

1	Cytoplasm
2	Enclave of cytoplasm
3	Plastid stroma
4	Plastid envelope
5	Plastoglobuli

2.47 Epidermal leucoplast from the green leaf of *Glechoma hederacea* (ground ivy). In the peripherally located stroma the thylakoid system is only poorly developed but several small starch grains (1) are evident. The centre of the plastid is occupied by a large vacuole (2). Cytoplasm (3), cell wall (4). (G-Os, TEM x 19,500.)

1	Starch grains
2	Vacuole
3	Cytoplasm
4	Cell wall

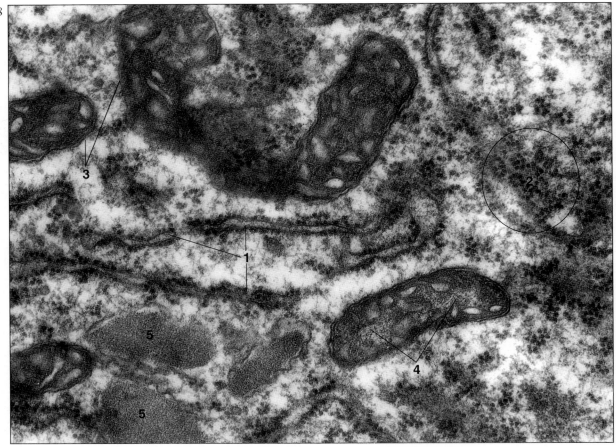

2.48 Densely cytoplasmic region from *Zea mays* (maize) root tip. Note the rough endoplasmic reticulum cisternae (1) covered with ribosomes which are often aggregated into polysomes (2). The mitochondria are enclosed within an envelope (3) and numerous irregular cristae lie within the matrix. These are connected with the inner membrane of the envelope; note also the fibrillar nucleoid zones (4) in the mitochrondrial matrix. Several sloughed off hypertrophied dictyosome cisternae (5) are also visible. (G-Os, TEM x 88,500.)

1	Endoplasmic reticulum cisternae
2	Polysomes
3	Envelope of mitochondrion
4	Fibrillar nucleoid zones
5	Hypertrophied dictyosome cisternae

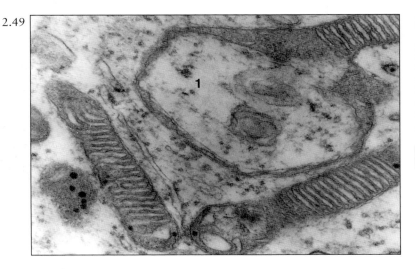

2.49 Mitochondria from callus tissue of *Taraxacum officinale* (dandelion). These probably represent segments of a single complex, polymorphic mitochodrion. The well-developed scalariform cristae are connected laterally with the inner membrane of the mitochondrial envelope; note also the large cytoplasmic enclave (1) in the upper mitochondrion. (G-Os, TEM x 53,500.)

1	Cytoplasmic enclave

2.50 Concentric cisternae of rough endoplasmic reticulum from the root tip of *Allium cepa* (onion). The cisternal membranes are densely studded with free ribosomes which are also packed free in the adjacent cytoplasm (1). The outer cisternal lamellae are inflated and contain fibrillar deposits (2) which probably represent protein synthesised on the associated ribosomes. (G-Os, TEM x 52,500.)

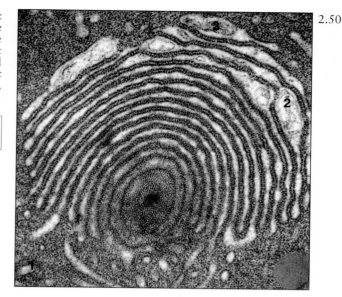

1 Cytoplasm
2 Fibrillar deposits

2.51 Dictyosomes from *Taraxacum officinale* (dandelion) callus sectioned in face view. Note that the cisternae are composed of anastomosing membranous tubules and that vesicles are forming at the margins. (G-Os, TEM x 36,500.)

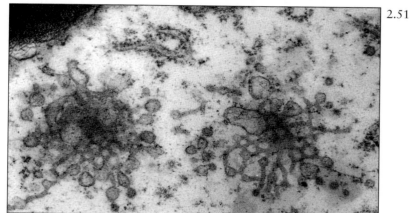

2.52 Peripheral cytoplasm of a cell from *Polytrichum commune* (hair moss). Note the parallel rough endoplasmic reticulum cisternae (1), numerous free ribosomes in the cytoplasm and a prominent dictyosome. This shows some polarity, with the overall thickness of the cisternae progressively increasing from forming face (2) to mature face (3); in the latter cisternae a distinct lumen is apparent within the investing membranes. Cell wall (4), plasmalemma (5). (G-Os, TEM x 43,000.)

1 Endoplasmic reticulum cisternae
2 Forming face of dictyosome
3 Mature face of dictyosome
4 Cell wall
5 Plasmalemma

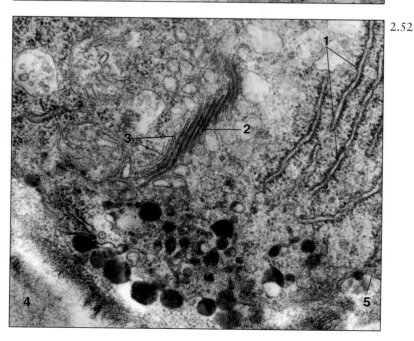

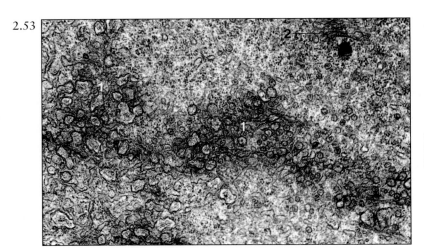

2.53 Obliquely sectioned cell plate forming between two daughter cells in *Linum usitatissimum* (flax). The cell plate forms from the abundant inflated vesicles (1) secreted by the dictyosomes (2). The vesicles migrate to the equator of the mitotic spindle and fuse with each other; thus the dictyosome membranes give rise to the plasmalemma. (G-Os, TEM x 23,000.)

1	Dictyosome vesicles
2	Dictyosome

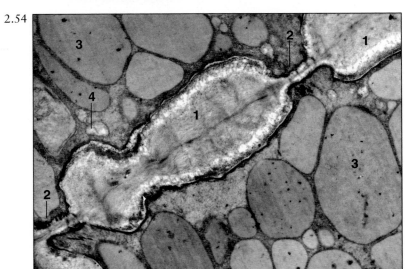

2.54 Partly hydrated cotyledonary storage cell from the germinating seed of *Phaseolus vulgaris* (bean). The greatly thickened wall (1) contain storage polysaccharides whilst the thinner regions represent simple pits (2) containing plasmadesmata. In the cytoplasm numerous large protein bodies (3) occur and several small mitochondria (4) can be distinguished. (G-Os, TEM x 6,500.)

1	Cell wall
2	Pits containing plasmadesmata
3	Protein bodies
4	Mitochondrion

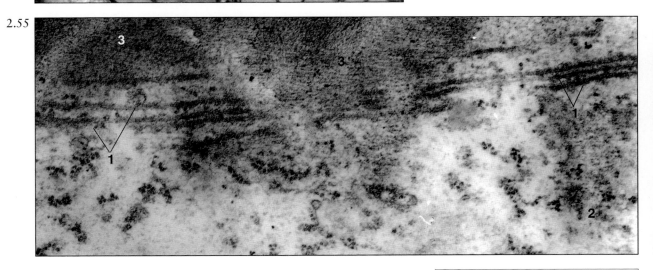

2.55 Interface between the cell wall and cytoplasm in *Pisum sativum* (pea) root. Note the group of longitudinally-sectioned, parallel microtubules (1) in the cytoplasm (2) and the similarly orientated microfibrils in the adjacent cell wall (3). (G-Os, TEM x 47,000.)

1	Microtubules
2	Cytoplasm
3	Microfibils in cell wall

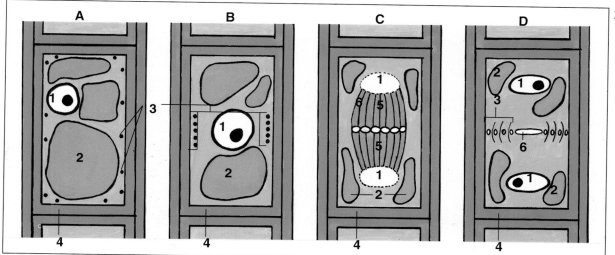

2.56 Diagrams showing the division of a parenchyma cell. **A,** interphase nucleus (1) located in the peripheral cytoplasm among large vacuoles (2) whilst the microtubules (3) lie adjacent to the cell wall (4). **B,** the nucleus (1) migrates to the central cytoplasm and the microtubules (3) become concentrated into a pre-prophase equatorial band just within the cell wall (4).

C, at the end of mitosis the envelopes (dotted) of the two progeny nuclei (1) are reconstituted at the poles of the mitotic spindle while at its equator dictyosome vesicles are fusing to form the cell plate. Peripheral microtubules are no longer present but the spindle 'fibres' (5) are composed of bundles of microtubules.

D, the progeny nuclei (1) are now fully formed, the spindle fibres have dispersed but a cell plate (6) has developed at the former equator. The plate spreads centrifugally and in the phragmosome at its margins short phragmoplast microtubules (3) lie interspersed with anatomosing dictyosome vesicles. At a slightly later stage than illustrated, the cell plate joins to the mother cell wall and two daughter cells are formed.

1	Nucleus
2	Vacuoles
3	Microtubules
4	Cell wall
5	Spindle fibres
6	Cell plate

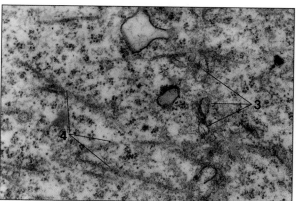

2.57, 2.58 LS of a dividing *Pisum sativum* (pea) root cell. **2.57** shows the centrifugally advancing cell plate (1) nearing the mother cell wall (2). **2.58** illustrates a detail of the beaded cell plate forming from the fusion of discrete dictyosome vesicles (3); note also the phragmoplast microtubules (4). (G-Os, TEM **2.57** x 4,900, **2.58** x 24,000.)

1	Cell plate	3	Dictyosome vesicles
2	Mother cell wall	4	Phragmoplast microtubules

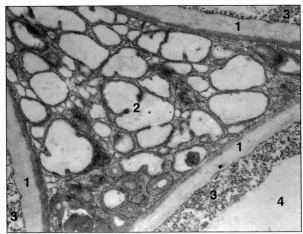

2.59 Development of an intercellular space in the cortex of *Pisum sativum* (pea) root. Three polygonal parenchyma cells have been sectioned at the angle where they interconnect and their thin cell walls (1) are separated by a greatly expanded middle lamella in which the matrix is breaking down to form a large cavity (2). Cytoplasm (3), vacuole (4). (G-Os, TEM x 10,500.)

1	Cell wall
2	Cavity
3	Cytoplasm
4	Vacuole

2.60

2.60 Diagrammatic representation of the wall structure of a fibre seen in transverse (bottom) and three-dimensional view (top). A thick secondary wall (1–3) surrounds the dead lumen (4). In the S1 and S3 layers of the secondary wall (1 and 3) the lamellae of parallel-orientated cellulose microfibrils are predominantly transversely orientated, but in the S2 layer (2) the microfibrils lie nearly vertical.

In the primary wall (5) the cellulose microfibrils are less abundant than in the secondary wall and tend to lie haphazardly. The primary walls of adjacent fibres are separated by a thin middle lamella (6) from which cellulose is absent. No pits are indicated although a few simple pits normally connect adjacent fibres.

1–3	Secondary wall:
1	S1 layer
2	S2 layer
3	S3 layer
4	Lumen
5	Primary wall
6	Middle lamella

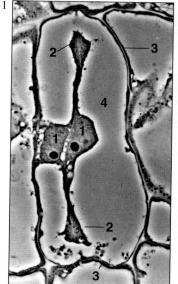

2.61 LS of a highly vacuolated parenchyma cell from the root of *Pisum sativum* (pea). The cell has recently undergone mitosis and the two progeny nuclei (1) are separated by a thin cell plate. This is covered on either side by a thin layer of cytoplasm which is distended at the margins of the plate to form a phragmosome (2). Mother cell wall (3), vacuole (4). (G-Os, Phase contrast LM x 925.)

2.62 Storage cell from a callusing cotyledon of *Phaseolus vulgaris* (bean). The thick mother cell wall (1) is connected to the thinner, tortuous, freely-forming wall (2) which develops *in vitro*. Note also the large vacuoles (3) which develop after hydrolysis of the protein bodies. (G-Os,TEM x 2,000.)

1	Mother cell wall
2	Freely-forming wall
3	Vacuoles

1	Nuclei
2	Phragmosome
3	Mother cell wall
4	Vacuole

2.63 Transfer cells from the stem of *Linum usitatissimum* (flax). The lower cell shows numerous wall profiles (1) which represent blindly-ending invaginations of the main wall (2) into the dense cytoplasm. The fibrillar darkly staining cores of these invaginations are surrounded by relatively translucent areas (possibly callosic) separated from the cytoplasm by the plasmalemma (3). (G-Os, TEM x 33,500.)

1	Wall invaginations
2	Main wall
3	Plasmalemma

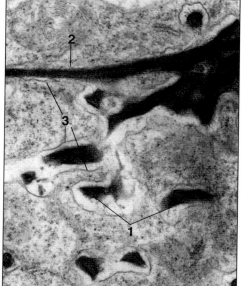

2.64

2.64 Pit field seen in face view of the plasmalemma of *Lycopersicon esculentum* (tomato). Note the numerous transversely fractured plasmodesmata (1) while the surface of the plasmalemma shows a number of small, probably proteinaceous, particles (cf., **2.17**). The linear structures are thought to represent evaginations of the plasmalemma (2) caused by tubular endoplasmic reticulum adjacent to the under (cytoplasmic) surface of the plasmalemma. View from the direction of shadowing (thick arrow). (FE, TEM x 44,000.)

1	Plasmodesmata
2	Evaginations of plasmalemma

2.65

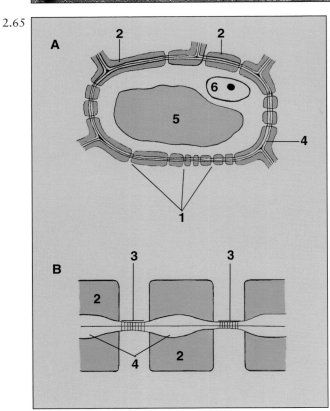

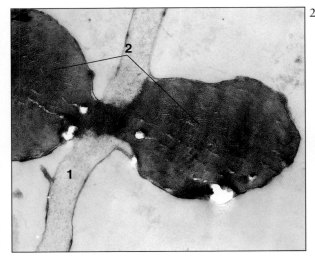

2.66

2.66 LS of an annular protoxylem element from the root of *Taraxacum officinale* (dandelion). Note the narrow 'holey' primary wall (1) and the densely-staining, lignified secondary wall thickenings (2). (G-Os, TEM x 19,000.)

1	Primary wall
2	Secondary wall thickenings

2.65 Diagrams of a xylem parenchyma cell. **A** illustrates numerous simple pits (1) in a thick secondary wall (2). **B** shows that the secondary wall (2) is not deposited at the pit fields (3, transverse lines represent plasmodesmata) and a uniformly-diametered simple pit is formed (cf., **2.54**). Primary wall (4), vacuole (5), nucleus (6).

1	Simple pits	4	Primary wall
2	Secondary wall	5	Vacuole
3	Pit fields	6	Nucleus

2.67 Development of a sieve tube and its companion cells in a flowering plant. **A**, sieve element precursor cell; note its thin primary wall (1), large central vacuole (2) and two nuclei (3) with a developing cell plate (4) between them. **B**, mature sieve element and companion cells. The largest cell formed from the precursor gives rise to the sieve element; which is enucleate but still retains its plasmalemma, modified plastids (5), mitochondria, endoplasmic reticulum and proteinaceous fibrils (6).

The end walls of the element now form the sieve plates and are perforated by pores which are often apparently occluded by fibrils. The cell wall is thickened at the sieve plate, and around the pores it contains amorphous callose. The two densely cytoplasmic, nucleated companion cells are formed after the further division of the smaller derivative of the precursor cell shown in **A**. Companion cell nucleus (7).

1	Primary wall
2	Central vacuole
3	Nuclei
4	Developing cell plate
5	Plastid
6	Proteinaceous fibrils
7	Companion cell nuclei

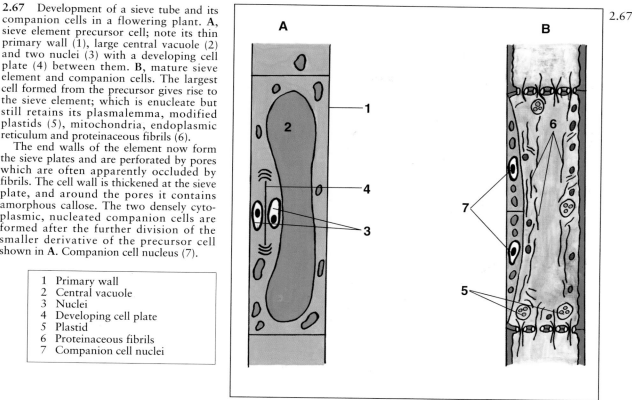

2.67

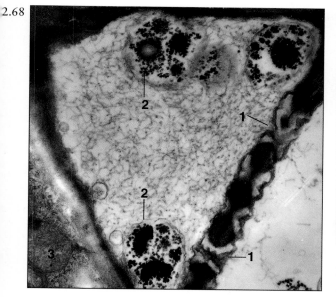

2.68 Sieve tube from the stem of *Linum usitatissimum* (flax). Note the obliquely inclined sieve plate with whitish regions of callose surrounding the sieve pores (1). Several plastids (2) containing starch are visible in the sieve tube and numerous fibrils of P-protein fill its lumen. Companion cell 3. (G-Os, TEM x 17,000.)

1	Sieve pores	3	Companion cell
2	Plastids		

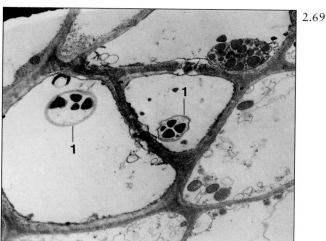

2.69

2.69 TS of secondary phloem (formed from a secondary thickening meristem) from the stem of the monocotyledon, *Dracaena*. Note the oblique sieve plates with callose (white areas) around the pores. Plastids (1) with starch grains are present in the lumina but P-protein is absent. (G-Os, TEM x 6,000.)

1	Plastids

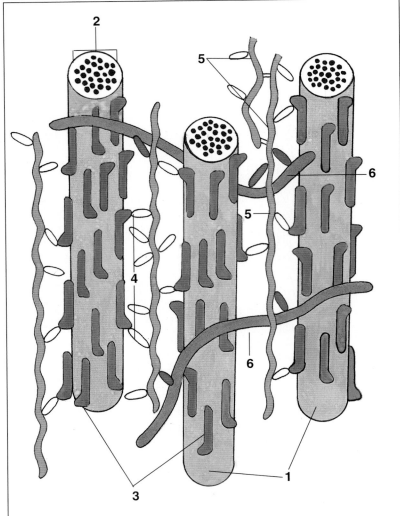

2.70 Model of the substructure of a primary cell wall. Each cellulose microfibril (1) contains up to 70 very long chains (2) of glucose monomers linked lengthwise by ß-1,4 bonds. The chains are hydrogen-bonded to neighbouring chains and the resultant microfibril is up to 30 nm wide and 5 μm long (cf., **2.17**). The wall contains large quantities of other polysaccharides, plus low levels of protein, which form the matrix. The hemicelluloses (3) are connected to the surface of the microfibrils by hydrogen bonding. Some hemicelluloses are linked, via neutral pectins (4), to acidic pectins (5). The glycoproteins (6) are probably connected to the acidic pectins.

1	Cellulose microfibrils
2	Glucose monomer chains
3	Hemicelluloses
4	Neutral pectins
5	Acidic pectins
6	Glycoproteins

CHAPTER 3

Plant histology

Distribution of cells and tissues

The vascular plant is composed of dermal, ground and vascular tissue systems (**3.1** to **3.4**). The phloem and xylem constitute the vascular system; both are complex tissues in which the conducting elements (**2.10, 2.68, 2.69**) are associated with other cell types (**1.4, 1.11, 3.5**). The ground system surrounds the vascular tissues (**3.1** to **3.4**) and comprises parenchyma, collenchyma and sclerenchyma (**3.6** to **3.10**). The dermal system is initially represented by the epidermis (**3.2, 3.3**) containing various cell types (**3.3, 3.11, 3.12**). Its structure is considered in Chapter 5.

In plants which undergo woody thickening, the epidermis typically becomes replaced by the cork and associated tissues (the periderm, **3.13**) whose structure is discussed in Chapter 6. Secretory cells do not normally develop as distinct tissues (except, for example, in nectaries) but rather occur either as surface structures (**3.11, 3.12**) or within other tissues (**3.4**).

Parenchyma

These cells are generally thin-walled and have large vacuoles (**2.24, 3.3, 3.6**). They form a continuous tissue in the cortex and pith (**1.27, 3.4**) and in the leaf mesophyll (**3.3**), while vertical strands and horizontal rays of parenchyma occur in the vascular system (**1.3, 3.14**). The leaves and stem of seedlings and small herbs are largely held erect by the collective turgor of the parenchyma cells (**3.3**) and the shoot wilts if insufficient water is available from the root system to replace transpiration losses (**1.12**).

Parenchyma cells are often polyhedral or elongate (**3.3, 3.6**) but stellate and irregular forms occur (**3.3, 3.15**). The mesophyll cell of the leaf is specialised for photosynthesis and contains numerous chloroplasts (**3.3**) while starch is frequently stored in the parenchyma of the root and stem (**1.14**). In plants growing in marshy conditions and in aquatics, the parenchyma of the shoot and root sometimes develops extensive intercellular spaces and is termed aerenchyma (**3.15**).

Mature parenchyma cells of the stem and root may resume division (**3.16**) and undergo partial dedifferentiation to form the lateral meristems of the phellogen and vascular cambium (**3.5**) from which the cork and secondary vascular tissues originate (**3.13, 3.17**). In some situations the parenchyma cells revert to an undifferentiated state (as found in the apical meristems, **2.2**) and give rise to new plant organs. This phenomenon results in the formation of lateral and adventitious roots (**3.18**) and in the development of primordia (**3.19**) which give rise to adventitious shoots on various parts of the plant (**3.20, 3.21**).

The development of adventitious organs is sometimes related to injury (**3.18**); the plant's regenerative capacity is exploited in horticulture by the rooting of shoot cuttings (**3.18**) and regeneration from various excised plant organs (**3.19** to **3.21**). Parenchyma cells should therefore be regarded as potentially totipotent and single cells derived from carrot and tobacco have given rise to completely new plants when cultured *in vitro* on a suitable nutrient medium with exogenous growth substances.

The parenchyma cell wall is often thin, with its growth normally terminating at the end of cell vacuolation (**2.24, 3.3, 3.15**). However, in seed storage tissue the walls may be greatly thickened (*Coffea, Phoenix, Phaseolus,* **2.54**) by non-cellulosic reserve carbohydrates, while parenchyma cells often develop thick lignified secondary walls (**1.3, 2.65**) in the xylem and pith. In transfer cells labyrinthine ingrowths of the wall (**2.63**) vastly increase the plasmalemmal surface, while simpler vertical ingrowths of the wall also occur in the plicate mesophyll of several conifers (**3.22**).

Collenchyma

These living cells are thick-walled but relatively pliable (**3.7**) and are located in the outer ground tissue (**3.23**). Collenchyma is of great value in the mechanical support of the young shoot but is rare in the root. The protoplasts of such cells are generally vacuolate (**3.7**) while their highly hydrated primary walls are unevenly thickened (**3.7, 3.24**) and contain large amounts of pectins and hemicelluloses. Collenchyma cells are usually elongate (**3.7**) and in transverse section often appear angular with thickening localised in the cell corners (**3.7, 3.24**) but the thickening may be confined to the tangential walls (**3.7**). Intercellular spaces sometimes occur in collenchyma and in older cells the walls may become lignified.

Sclerenchyma

This tissue is characterised by its thick, normally lignified, secondary walls and the protoplasts are usually dead. Sclerenchyma is the principal mechanical tissue of the non-secondary thickened plant organs (**3.25, 3.26**) and exists as either sclereids (**3.8, 3.9, 3.25**) or fibres (**1.4, 3.1, 3.10, 3.26, 3.27**).

Sclereids frequently occur either singly (**3.8A**) or in small aggregates (**3.9**) and may develop in the epidermis as well as internally in the plant. They vary considerably from more or less isodiometric stone cells with prominent simple pits (**3.9**), to branched osteosclereids with tapering branches (**3.8, 3.25**). Sclereids occur in leaves of some plants (**3.25**), in the hardened fruit walls of nuts and stone fruits and in the seed coats of many legumes.

Fibres commonly occur in groups, forming strands (**3.2, 3.10, 3.26, 3.27**); the individual cells are slender and highly elongate, with their tapering end walls overlapping and simple pits linking adjacent fibres (**3.10**). Their walls show extensive secondary deposition (**2.7, 2.60, 3.26**) and their lumina are generally much narrower than those of xylem tracheary elements (**3.27, 3.28**). The walls of mature fibres are generally lignified, hard and incapable of extension so that they mature in regions of the plant in which elongation has ceased (**3.26, 3.27**).

In some dicotyledons the fibres are divided by several unthickened cross walls; these are designated as septate fibres and often retain their protoplasts. The thin-walled tips of fibres frequently continue to elongate after their mid regions have formed extensive secondary walls (**1.4, 2.7, 3.2**). In *Linum* (flax) the individual fibres show intrusive growth between neighbouring parenchyma cells (**3.29**) and may reach 6 cm in length.

Textile fibres are obtained from the extensive strands of primary phloem fibres that occur in the stems of several dicotyledon (e.g. *Boehmeria, Corchorus*) and are fairly flexible, since they are usually only moderately lignified or non-lignified (e.g. *Linum*, **1.4, 3.2**). However, in fibres from monocotyledonous leaves (e.g. *Agave, Phormium, Sansevieria,* **3.26**) the lignification is greater and the extracted fibres are coarse and stiff.

Secretory tissues

Secretory trichomes (hairs) and glands (see Chapter 5) often develop in the epidermal and sub-epidermal tissues. In the insectivorous plant *Drosera* the leaves are covered by long and complex multicellular hairs (**3.11, 3.12**). Their glandular heads are coated with a viscous secretion containing digestive enzymes in which insects become trapped. This secretion originates in the peripheral layer of densely cytoplasmic secretory cells in the glandular head and migrates to the surface via numerous pores in the cuticle. The gland contains a central file of tracheids which is separated from the secretory cells by an endodermis with Casparian bands of the type present in the young root (see Chapter 4).

Nectaries are glands which secrete a sugary solution (nectar); they are located superficially and occur either on the flower (floral nectaries) or on various parts of the shoot (extra-floral nectaries). In the inflorescence of *Euphorbia* (**3.30**) a connate bract bears four oval nectaries. The several layers of secretory cells are densely cytoplasmic and the nectaries are supplied with vascular tissue. A shallow layer of nectar is secreted, so attracting various flies which effect pollination.

Hydathodes frequently occur on leaves and excrete water (**3.32**) from the leaf margins and tips. The hydathode consists of modified mesophyll tissue into which the water is discharged from tracheids. The enclosing epidermis bears stomata which remain permanently open and through which water is secreted. This guttation may be a mechanism to protect shoot tissues from becoming waterlogged in situations in which root pressure is excessive. However, many hydathodes contain transfer cells, indicating that active excretion of water occurs at the same time as minerals are unloaded within the tissue and become available for leaf growth.

A large variety of secretory structures also occurs internally in the plant body (**3.4, 3.17**). In *Pinus* and other conifers resin canals (**3.17, 3.22**) are formed schizogenously by the separation of adjacent cells from each other to form a central duct. Resin, which contains various terpenes, is secreted into the duct by the lining epithelial cells. Mucilage ducts (**3.4, 3.31**) are also formed schizogenously. Laticifers produce a milky secretion termed latex (**3.33**) which probably represents a deposit of various metabolic byproducts produced by the plant.

Laticifers occur in about 900 genera of angiosperms and are often branched and usually extend throughout the plant body in various tissues. In the fleshy, largely parenchymatous, root of *Taraxacum* the secondary phloem is especially well supplied with laticifers (**3.34**). These are closely associated with the strands of sieve tubes (**3.35**) which occur in concentric rings isolated by intervening phloem parenchyma (**3.34**).

The articulated laticifers of *Taraxacum, Hevea* (the main source of commercial rubber) and many other genera originate from the breakdown of the intervening walls between contiguous cells. However, non-articulated laticifers form from a single cell which often becomes multinucleate. Both types of laticifers may branch and anastomose (3.36).

Phloem

This is a complex tissue composed of sieve elements, companion cells, parenchyma and sclerenchyma (1.4, 1.11, 3.37), whilst laticifers sometimes also occur (3.35). The conductive sieve elements (2.68, 2.69) of the phloem generally function for only a few months and in the protophloem (2.3) are evanescent. In *Vitis* and *Tilia* (3.37, 3.38) they function over several seasons, while in palms they apparently translocate for many years.

The angiosperm sieve tube is composed of sieve tube members joined end to end at their sieve plates (2.67, 3.39). The individual members are 50 to 150 µm long and up to 40 µm wide and have primary walls, although these may be thickened. The sieve plates separating sieve tube members are perforated by numerous pores (2.6). On the transversely situated sieve plates of *Cucurbita* (3.40) the pores may reach 15 µm in diameter. In species with obliquely-inclined end walls, the sieve plates are compound (3.42) and are composed of several sieve areas with small pores only one to several micrometres wide.

The sieve tube member represents a living, but highly modified, enucleate cell (2.67). During its maturation the tonoplast and most organelles are lost (2.67 to 2.69). The plasmodesmata develop into the vastly enlarged sieve pores (2.6, 3.40), but the plasmalemma remains intact and lines the margins of the pores and wall (2.6). On the longitudinal walls sieve areas may occur (3.42) but these are less well-defined than the sieve plates and the pores are smaller.

In actively translocating sieve tubes the pores are generally considered to be open (3.43). However, in most sections prepared from dicotyledonous material the pores are blocked by plugs of P-protein (2.6, 2.68, 3.39) while callose deposition at the margins of the pores (2.68) greatly reduces their diameter. Similar sealing of sieve tubes is thought to occur in damaged tissue on the intact plant. In monocotyledons (2.69) P-protein is rare and it is apparently absent in gymnosperms. In non-flowering vascular plants the translocating elements are discrete sieve cells; these are commonly very elongated with their small sieve areas distributed over both the vertical and tapering end walls.

Sieve tubes are typically associated with both parenchyma cells and more densely cytoplasmic companion cells. Sieve tube members and companion cells arise from common precursor cells (2.67) and plasmodesmatal connections between them are abundant, with those in the companion cell wall usually branched (3.43). Companion cells apparently supply ATP to the sieve tubes; in the minor veins of the leaf they function, along with parenchyma, as intermediary cells in the accumulation and loading of photosynthates into the sieve tubes (3.43).

The pressure-flow hypothesis of translocation (3.43) suggests that sugars and other nutrients are loaded by molecular pumps into the sieve tubes of the leaf. This generates a high osmotic pressure in the tubes and consequently water is absorbed. The increased turgor pressure causes flow from one element to another via the open sieve pores. In growing regions of the plant (2.3), and storage regions (1.14), the sugars are pumped out from the sieve tubes into the adjacent tissues and water follows, so that a mass flow of nutrients is established from source to sink (3.43). Translocation in the phloem allows fast movement of nutrients (usually about 1 metre per hour) and the sap contains up to 250 mg/litre of sugar plus other nutrients and plant hormones.

In secondary-thickened plant organs the older phloem has additional roles. It is a principal component of the protective bark of trees (3.13) and in small shoots or twigs the phloem fibres provide considerable mechanical support (3.37). Secondary phloem parenchyma (3.37) provides an important storage tissue. When the buds break in deciduous woody species, large quantities of carbohydrates and nitrogenous substances are mobilised in the phloem parenchyma. The soluble products are then transported to the expanding new leaves to sustain them before they are fully photosynthetic. Secretory tissues often occur within the phloem and in *Hevea* (rubber tree) the laticifers in the phloem secrete various polyterpenes; the milky liquid which exudes when the tree is tapped is refined to provide rubber.

To accommodate the increased circumference of the growing root or stem, the outermost secondary phloem expands laterally by the resumed division and growth of its parenchyma. This expansion is commonly obvious in the ray tissue which may flare outwards from the vascular cambium towards the non-functional outer phloem (3.37). In most woody species the first-formed cork cambium (phellogen) becomes non-functional; new cambia arise progressively more deeply internally and eventually from the parenchyma of secondary phloem, where they often appear in cross section as discontinuous but overlapping layers (3.41). This older phloem becomes sloughed off in the successively formed layers of the peeling bark (3.13).

Xylem

This complex tissue has two principal roles: the transport of large quantities of water from the root to the shoot in the tracheary elements (**1.12**) and the mechanical support of the aerial plant body (**3.44**). This support is provided both by the tracheary elements and the associated (often thicker-walled) non-conducting fibres (**1.3, 3.28**). Additionally, in the secondary xylem the axial and ray parenchyma (**1.3, 3.14, 3.45**) store food and water.

Mature tracheary elements (**1.11, 3.17**) are dead and have lost their protoplasts (**2.9, 2.10**). Their secondary walls are thickened relative to the primary walls (**2.8, 2.10**) and, due to lignification, are impermeable except at the pits where only primary wall is present (**2.10**). The tracheary elements are elongated and water moves along their lumina from the root to the shoot in the transpiration stream (**1.12**). The absence of a plasmalemma allows the water to pass fairly freely from one element to another via the numerous pits (**3.46**).

In nearly all angiosperms the tracheary elements comprise both tracheids and vessels (**3.47**) but generally only tracheids occur in gymnosperms and lower vascular plants (**3.17, 3.27, 3.48**). A tracheid is derived from a single cell and has no perforations; it is elongated, with tapering ends (**3.47**) and in conifers the tracheids (**3.49**) may reach a centimetre in length. By contrast, vessels are composed of a tube-like series of two to many vessel elements lying end to end. They are directly linked through their perforation plates (**3.47, 3.50**) which represent the remnants of their original end walls (**3.51**). Vessel elements tend to be shorter but wider than tracheids (**3.47**).

The perforation plate shows either a single large pore (**3.47**) or in compound plates a number of elongated pores which are commonly scalariform (**3.14, 3.47**). Because of these open pores vessels generally show a lower resistance to water movement than tracheids, where the closed pits impede water flow. In ringporous wood (**1.3**), the wide-diametered vessels (**3.28**) apparently extend many metres along the tree trunk. The tips of vessels are imperforate but the numerous pits allow water to move into adjacent tracheary elements.

The extent and type of the pitting in tracheary elements is variable (**3.47 to 3.52**). The protoxylem in the shoot shows secondary wall deposition of an annular or helical pattern (**2.8, 2.9, 3.50**). The primary wall between thickenings becomes greatly extended after the protoxylem element dies whilst its non-cellulosic components are digested. The stretched wall often appears 'holey' under TEM (**2.66**) and may rupture leaving a protoxylem cavity (**1.11**). In non-elonga-ting regions of the plant, metaxylem elements show much more extensive secondary wall deposition of various patterns (**3.52**).

In a scalariform element at least half of the primary wall is covered by secondary wall and the pits are horizontally elongated and usually bordered (**3.52**). In reticulate elements the thickening is more irregular (**3.52**) whilst in pitted elements (**3.47, 3.48, 3.52**) a greater proportion of the wall is secondary. Their pits occur in horizontal rows (opposite pitting, **3.52**) or diagonally (alternate pitting **3.49, 3.57**).

In these various modes of secondary wall thickenings the intervening pits are bordered (**2.10**). In conifer tracheids the centre of the pit membrane (torus, **3.48**) is thickened and lignified, but the periphery (margo) has only a loose cellulose network and is permeable. The pits between adjacent tracheary elements are abundant and bordered (**2.9, 2.10, 3.46**), but there are few connections to fibres. The pits which link with parenchyma cells are either simple or half-bordered on the tracheary element side (**3.46**).

Structure of wood

In woody plants the formation of secondary vascular tissue is typically periodic, since the vascular cambium becomes dormant in unfavourable environmental conditions. This typically results in the formation of growth rings in the secondary xylem of the tree (**3.44, 3.45**). In non-tropical species the cambial activity is limited by temperature and the rings usually represent annual increments (**1.3, 3.17, 3.41, 3.44, 3.45**). Generally the last-formed layers of xylem in a growth ring are composed of narrower cells with thicker walls than the earlier wood (**3.53**), so that growth rings are often visible to the naked eye (**3.44**). In many tropical trees (**3.54**) and desert succulents, growth rings are not obvious.

The newly-formed tracheary elements of the sapwood conduct water for a relatively short time. In stressful environments (for example nutrient deficient conditions) they often cease to function by the end of the first year, but in tropical trees they may remain active longer. The cessation of water transport in a tracheary element results from cavitation (**3.46**). The tension on the water columns within tracheary lumina is greatest in the widest elements.

However, when water is plentifully available in the soil, most conduction occurs within them since they offer less resistance to water flow than narrow elements. When the water supply is restricted, the tension on water columns in wide elements may become very severe. If a column breaks, an embolism quickly expands within the tracheary element but is blocked off by the pit membranes and does not spread to adjacent tracheary elements (**3.46**).

The non-conducting tracheary elements in the heartwood of a tree (**3.44**) constitute the great bulk of the wood and provide the tree's principal support. Vessels in the heartwood frequently become filled by parenchymatous tyloses (**1.3**, **3.55**), which intrude via the pits from adjacent parenchyma cells. Gums and various polyphenols are often deposited when tyloses are absent. In heartwood the walls of parenchyma cells may become lignified and the reserve food and water is withdrawn, so that overall the wood becomes drier.

The secondary xylem of dicotyledons generally contains a high proportion of thick-walled fibres and are termed hardwoods (**1.3**, **3.14**, **3.28**, **3.56**), but *Ochroma* (balsa wood) has thin-walled fibres and extensive rays (**3.57**). Conifer woods generally lack fibres (**3.17**, **3.27**, **3.58**) and are designated softwoods. In conifers axial parenchyma is rare and food reserves are stored in the rays (**3.58**).

Radial movement of water in conifer wood is largely restricted to these rays since in the tracheids pits normally occur only on the radial walls (**3.48**, **3.53**). The wood of dicotyledons is usually more complex than in conifers and contains varying-diametered vessels, tracheids and fibres (**1.3**, **3.14**, **3.59** to **3.62**) while axial and ray parenchyma is normally abundant (**3.14**, **3.57**, **3.62**).

Rays generally consist of procumbent (somewhat radially elongated) parenchyma cells. However, in angiosperms the top, bottom and margins of the ray may also contain specialised upright (axially elongated) parenchyma. In some gymnosperms (**3.48**) the top and bottom of the ray is composed of dead and lignified ray tracheids whilst the ray parenchyma may also be lignified. Rays in gymnosperm wood are almost always one cell wide (uniseriate, **3.53**, **3.58**) but in dicotyledons both uniseriate and multiseriate rays (**3.62**, **3.63** to **3.65**) may occur.

Carbohydrates are often stored in large quantities in rays; in deciduous trees these are mobilised at the onset of the growing season and transported to the expanding buds. Sometimes, as in sugar maple (*Acer saccharum*), these solutes are conducted in the xylem elements and pass from the ray to the tracheary elements via the numerous pit connections in the upright ray parenchyma cells.

Angiosperm wood normally contains vessels, but in a few genera only tracheids occur (**3.65**, **3.66**). In ring porous trees (**1.3**, **3.67**), the new season's early growth contains large vessels but relatively few tracheids and fibres (*Catalpa, Fraxinus, Quercus, Robinia*). Later in the year the vessels become smaller while the proportion of fibres and tracheids in the wood increases. Diffuse porous species (**3.68**) show smaller pores and are more common than ring porous species. Large, isolated vessels sometimes occur in the wood and are associated with numerous tracheids (**3.67**), but more commonly the vessels are clustered in various patterns (**3.28**).

Although the main trunk of a tree normally grows more or less vertically the branches grow horizontally or obliquely, and their secondary xylem shows reaction wood with distinctive anatomical characteristics (**3.44**, **3.45**, **3.69**). In dicotyledons this is termed tension wood; it forms on the upper side of the branch and the growth ring is usually thicker here (**3.44**, **3.45**), with the xylem often developing unlignified gelatinous fibres. By contrast, in conifers the increased xylem formation on the lower side of branches (**3.69**) yields a brittle compression wood containing thicker-walled and heavily lignified tracheids.

3.1 TS of the petrified stem of the fossil seed fern *Lyginopteris*. In this Carboniferous plant the same tissue systems occur as in present-day flowering plants. The ground tissue is represented by a prominent pith (1) and cortex which shows an outer network of fibres (2). The dermal system and leaf bases have, however, not been preserved. The ring of secondary xylem (3, composed of radially seriated tracheids) and the poorly preserved phloem (4) represent the vascular system. A leaf trace (5) is also apparent.

1	Pith
2	Network of fibres
3	Secondary xylem
4	Phloem
5	Leaf trace

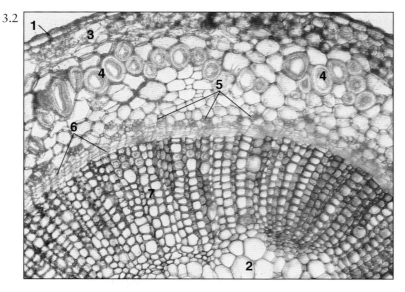

3.2 *Linum usitatissimum* (flax) showing the distribution of vascular, ground and dermal systems in the stem of a dicotyledon. The latter is composed of a single-layered epidermis (1) but the ground system comprises both the parenchymatous pith (2) and the cortex (3). The outer limit of the vascular system is marked by large phloem fibres (4) while internally groups of conducting elements (5) lie in the phloem parenchyma. A conspicuous vascular cambium (6) separates the phloem from the secondary xylem (7). (LM x 130.)

1	Epidermis
2	Parenchymatous pith
3	Cortex
4	Phloem fibres
5	Phloem conducting elements
6	Vascular cambium
7	Secondary xylem

3.3 TS of the lamina of a bifacial leaf of the dicotyledon *Glechoma hederacea* (ground ivy). The dermal system comprises the ad- and abaxial epidermis (1 and 2) with the stomata (3) confined to the abaxial surface. The chlorenchymatous ground tissue consists of a single palisade layer (4) and a thicker layer of spongy mesophyll (5). Numerous large chloroplasts are visible in the mesophyll but are absent from the epidermis. The vascular system is represented by the veinlet (6). (G-Os, LM x 330.)

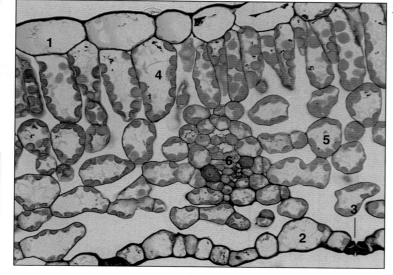

1	Adaxial epidermis
2	Abaxial epidermis
3	Stoma
4	Palisade layer
5	Spongy mesophyll
6	Veinlet

3.4 TS of the aerial root of *Monstera* (a monocotyledon). The central polyarch vascular system shows prominent wide-diametered vessels (1), while the ground tissue consists of a thick-walled, lignified pith (2) and a wide parenchymatous cortex (3). These are demarcated by the endodermis (4). Mucilage ducts (5). (LM x 85.)

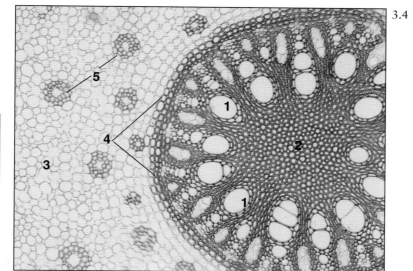

1	Vessels
2	Lignified pith
3	Parenchymatous cortex
4	Endodermis
5	Mucilage ducts

3.5 TS of the herbaceous stem of the dicotyledon *Zinnia* showing detail of a vascular bundle. This is enclosed by the parenchymatous ground tissue of the pith (1) and cortex (2). The xylem contains several mature tracheary elements (3) with wide lumens and thickened walls but these are interspersed by axial parenchyma (4). The vascular cambium (5) forms xylem centripetally and phloem centrifugally. The latter consists of axial parenchyma in which are interspersed densely staining companion cells (6) and apparently empty sieve tubes. (G-Os, LM x 330.)

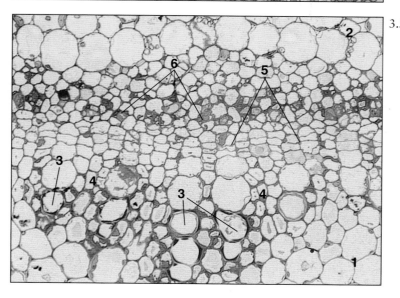

1	Pith
2	Cortex
3	Tracheary elements
4	Axial parenchyma
5	Vascular cambium
6	Companion cells

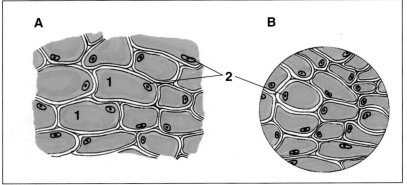

Diagrams of parenchyma cells viewed in longitudinal (**A**) and transverse (**B**) sections of a stem. Note the large vacuoles (1) and the peripheral cytoplasm containing the nuclei (2) and other organelles. Thin cellulosic primary walls enclose the protoplasts and small intercellular spaces occur at the angles of the cells where the middle lamellae are breaking down.

1 Vacuoles	2 Nuclei

3.7

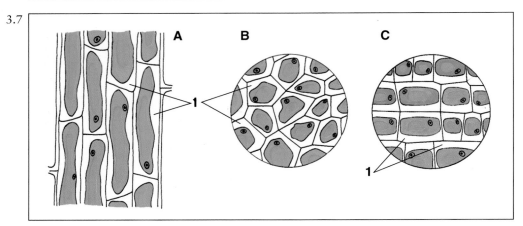

1 Primary walls

3.7 Diagrams of collenchyma cells in longitudinal (**A**) and transverse views (**B, C**) of a stem. Their protoplasts are similar to those of parenchyma cells (cf., **3.6A, B**) but the cellulosic primary walls (1) are unevenly thickened. Two common variants are illustrated: angular (**A** to **B**) and lamellar collenchyma (**C**) and in both intercellular spaces are absent.

3.8

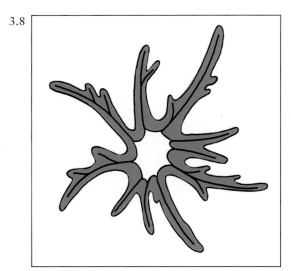

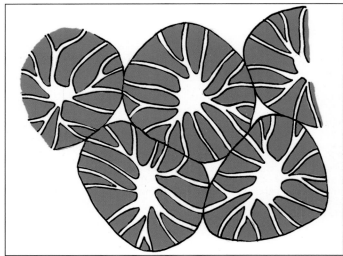

3.9

3.8, 3.9 Diagrams of sclerenchymatous elements with thick, lignified secondary walls. Mature cells are generally dead and prominent lumina replace their protoplasts. **3.8** Much-branched asterosclereid. **3.9** Group of isodiametric stone cells with branched simple pits.

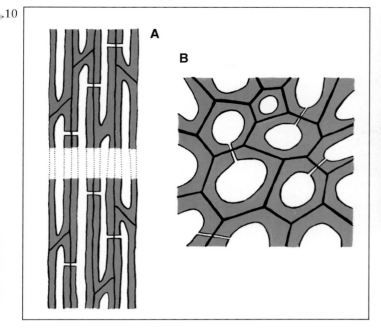

3.10 Diagrams of fibrous sclerenchymatous elements (with thick, lignified secondary walls) seen in (**A**) longitudinal and (**B**) transverse views. These highly elongate cells have tapering (often branched) tips and simple pits are often frequent in their walls. Fibres constitute a major mechanical support system in the shoot.

3.11 Leaf of the insectivorous dicotyledon *Drosera rotundifolia* (sundew cf., **3.12**). The crowded long hairs, with glandular heads sticky with secretions, are especially prominent on the adaxial laminal surface. These epidermal glands are complex multicellular structures with the cylindrical stalk containing a central tracheary strand.

The epidermal cells of the glandular head secrete a viscous fluid in which small insects become stuck and the adjacent hairs then bend towards the victim. Enzymes within the secretion digest the insect's tissues and the soluble products are absorbed by the gland and translocated to the leaf and elsewhere in the plant. (*Copyright of T. Norman Tait.*)

3.12 Flowering specimens of the insectivorous dicotyledon *Drosera rotundifolia* (sundew). These were growing on the pebble-strewn margin of a small lake in a very thin soil which was deficient in minerals, particularly nitrogen and phosphorus. Small insects become trapped in the mucilage secreted by numerous glands on the leaf surface (cf., **3.11**). The digestive enzymes in the secretion breakdown the proteins in the insect and provide the plant with an important sources of nitrogen and other minerals.

3.13 Stack of cork harvested from the dicotyledon *Quercus suber* (cork oak). This cork replaces the epidermis of the young stem and represents the dermal system of the tree. At intervals of about 10 years a layer several centimetres thick (consisting of cork formed by successive cork cambia) is removed from the tree but a thin layer of newly produced cork is left on the trunk to protect the phloem within. Commercial cork is highly water-resistant and provides excellent thermal insulation.

3.14

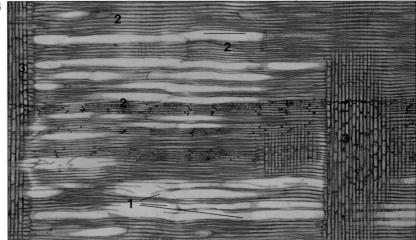

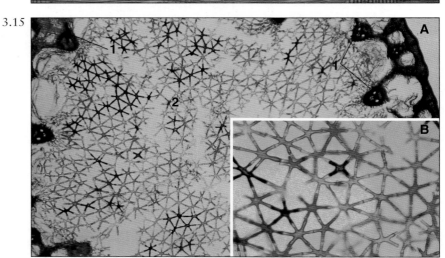

3.15

A

B

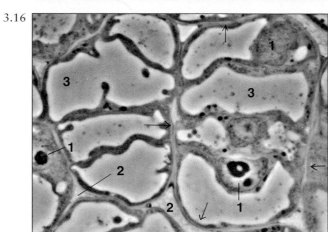

3.16

3.14 RLS through the diffuse porous wood of the dicotyledon *Magnolia grandiflora*. Note the large-diametered pores (vessels) with scalariform perforation plates (1) and the narrow-diametered fibres (2). Wide parenchymatous rays (3) are present and axial parenchyma is also evident. (LM x 35.)

1	Scalariform perforation plates
2	Fibres
3	Parenchymatous rays

3.15 TS of the hydrophytic stem of the monocotyledon *Juncus communis* (rush). **A**, although this is a monocotyledon, the vascular bundles (1) are peripherally distributed and most of the stem is occupied by an aerenchymatous pith (2). **B**, detail of the aerenchyma cells which connect, via stellate arms, with adjacent cells; note the enormous apoplastic system represented by the intercellular spaces and cell walls. (LM **A** x 35, **B** x 85.)

| 1 | Vascular bundles |
| 2 | Aerenchymatous pith |

3.16 Reactivated cortical parenchyma cells from an *in vitro* cultured *Pisum sativum* (pea) root. The mother parenchyma cell (with its wall demarcated by arrows) has divided into four smaller cells and three derivative nuclei (1) are visible. The other cell has similarly divided but no nuclei are visible. Intercellular space (2), vacuole (3). (G-Os, Phase LM x 1,150.)

1	Nucleus
2	Intercellular space
3	Vacuole

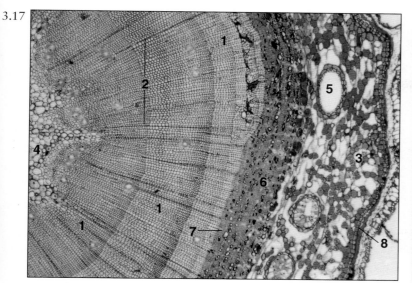

3.17 TS of a young stem of the gymnosperm *Pinus* (pine). Note the origin of the secondary vascular tissues and cork from the vascular and cork cambia respectively. The secondary xylem (1) shows several growth rings whose conducting elements consist of tracheids only, while numerous rays (2) traverse it. Cortex (3), pith (4), resin duct (5), secondary phloem (6), vascular cambium (7), cork cambium (8). (LM x 35.)

1	Secondary xylem	5	Resin duct
2	Rays	6	Secondary phloem
3	Cortex	7	Vascular cambium
4	Pith	8	Cork cambium

3.18 Stem tip cuttings of the dicotyledon *Kalanchoe* with numerous basal adventitious roots formed after growing for six weeks in compost. These roots originated endogenously from vascular parenchyma cells which as a result of division and dedifferentiation (cf., 3.16) eventually gave rise to new roots.

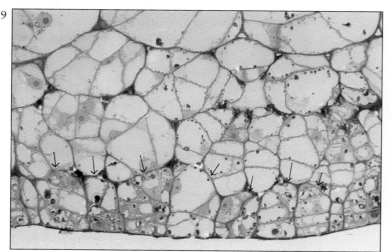

3.19 TS of the outer tissues of a mid-rib of a *Begonia rex* leaf segment cultured *in vitro* for several weeks. Note the repeated divisions in the epidermal cells (arrows indicate the inner limits of the original epidermis) and the larger hypodermal parenchyma. Later, numerous adventitious buds arise *de novo* from these meristematic cells. (G-Os, LM x 450.)

3.20 Leafy adventitious bud (1) arising *de novo* at the base of a parent leaf (2) of *Begonia rex* (a dicotyledon). This leaf was still attached to the parent plant although normally in this genus adventitious buds only develop on detached leaves.

1	Adventitious bud
2	Parent leaf

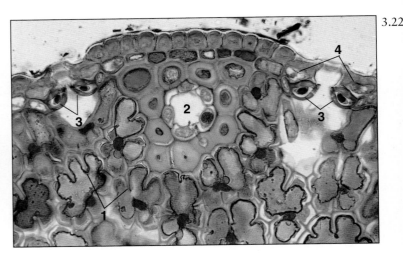

3.21 Non-sterile excised root segments of the dicotyledon *Armoracia rusticana* (horse radish) cultured for several weeks *in vitro*. Note the numerous leafy adventitious buds which arise from the cork cambium of the root; adventitious roots also arise from this tissue.

3.22 TS of a xeromorphic leaf of of the conifer *Pinus* (pine) showing the epidermis and mesophyll. The thin mesophyll cell walls possess numerous vertical ingrowths (1), but the protoplasts of these cells have become plasmolysed so that a gap separates them from the walls. Note the resin duct (2) in the mesophyll and also the guard cells (3) which are sunken beneath the subsidiary cells (4) in the epidermis. (LM x 110.)

1	Wall ingrowths	3	Guard cells
2	Resin duct	4	Subsidiary cells

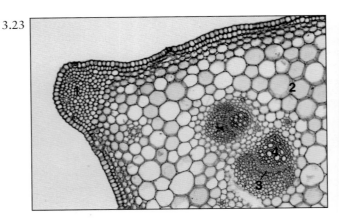

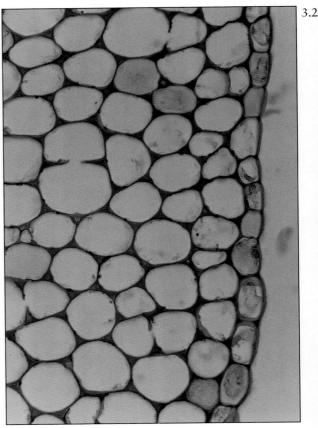

3.23 TS of the petiole of the dicotyledon *Sanicula europea* (sanicle) showing the peripheral location of collenchyma (1). This living tissue has thickened cellulosic walls and its location helps to support the young leaf. Parenchyma (2), primary phloem (3), primary xylem (4). (LM x 60.)

1	Collenchyma	3	Primary phloem
2	Parenchyma	4	Primary xylem

3.24 TS of the stem of *Coleus* (a dicotyledon) showing detail of the collenchyma. This peripheral tissue (cf., **3.23**) is of the angular form with additional cellulose thickening deposited at the angles of these cells (cf., **3.7B**). (LM x 335.)

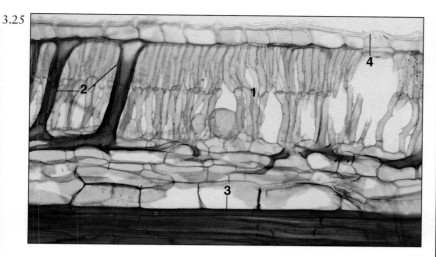

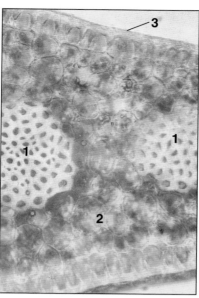

3.25 LS of the xeromorphic leaf of *Hakea* showing columnar sclereids. The palisade mesophyll (1) of this leaf is strengthened by sclereids (2) which have branched ends terminating beneath the epidermis and at the sheath of the vascular strand (3). Note the thick epidermal cuticle (4). (LM x 120.)

1	Palisade mesophyll	3	Vascular sheath
2	Sclereids	4	Cuticle

1	Lignified fibres	3	Cuticle
2	Mesophyll		

3.26 TS of the xeromorphic leaf of the monocotyledon *Sansevieria trifasciata* (bowstring hemp). Note the longitudinal strands of thick-walled, lignified fibres (1) in the mesophyll (2); these fibres are used commercially for cordage. A thick cuticle (3) covers the epidermal cells. (Fresh section, LM x 230.)).

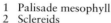
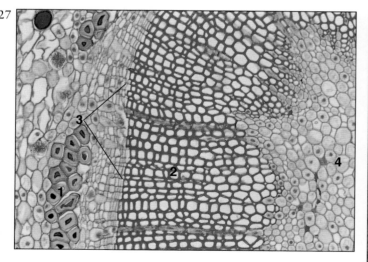

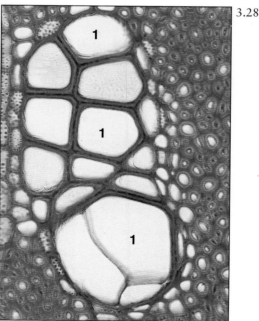

3.27 TS of a young twig of the gymnosperm *Ginkgo biloba* (maiden hair tree). Note the contrast between the thick-walled phloem fibres (1) and the thinner-walled tracheids of the secondary xylem (2). Vascular cambium (3), pith (4). (LM x 140.)

1	Phloem fibres	3	Vascular cambium
2	Secondary xylem	4	Pith

1	Large-diametered vessels

3.28 TS of the wood of the dicotyledon *Robinia pseudoacacia*. Note the cluster of large-diametered vessels (1) surrounded by much smaller fibres with very narrow lumina (cf., **1.3**). (LM x 260.)

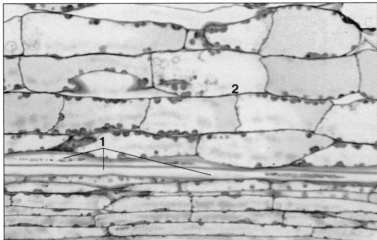

LS of the young stem of the dicotyledon *Linum usitatissimum* (flax) showing intrusive fibres. The thin-walled fibres (1) are growing between adjacent chlorenchyma cells (2). Flax fibres are generally unlignified and the multinucleate protoplasts are persistent. At maturity the fibres are extremely thick-walled (cf., **2.7**) and may be extracted to produce linen. (G-Os, LM x 285.)

| 1 | Fibre |
| 2 | Chlorenchyma cells |

3.30

3.30 Clustered inflorescences of the dicotyledon *Euphorbia cyparissias*. Each 'flower' is actually a small inflorescence with a pair of bracts (1) at its base. A single cup-shaped bract bears four yellow nectaries (2) and in the centre lies a prominent ovary (3) terminated by three styles. This represents the solitary female flower which lacks a perianth. At the base of the pedicel (4), which bears the female flower, lie the male flowers (5) with each represented by a single stamen.

1	Bracts
2	Nectaries
3	Ovary
4	Pedicel
5	Male flowers

3.31 TS of a cortical mucilage duct in the aerial root of the monocotyledon *Philodendron saggitifolium*. These occur in the parenchymatous cortex (1, cf., **3.4**) and each consists of a narrow central duct (2) surrounded by a single layer of secretory cells (3). The mucilage duct is itself enclosed withing several layers of fibres (4). (LM x 550.)

1 Parenchymatous cortex
2 Central duct
3 Secretory cells
4 Fibres

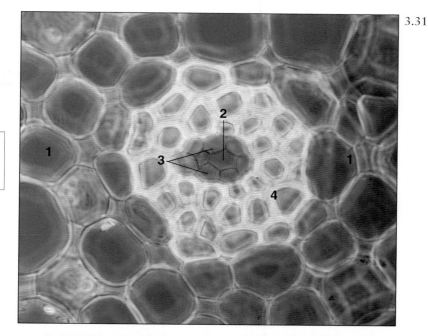

3.31

3.32 Leaf of the dicotyledon *Alchemilla* showing guttation. Water is excreted from numerous hydathodes (modified stomata) located at the leaf margins. Guttation may be a mechanism to protect the intercellular space system of the leaf from becoming waterlogged in conditions of excessive root pressure. (*Copyright of T. Norman Tait.*)

3.32

3.33 Xeromorphic shoot system of the dicotyledon *Euphorbia canariensis* with its leaves reduced to spines. The stem contains an elaborate laticiferous system and the wounded stem has exuded latex (1) which is congealing on to the volcanic rock on which the bush is growing.

1 Latex

3.33

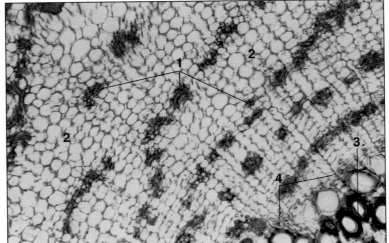

3.34

3.34 TS of the secondary phloem in the root of the dicotyledon *Taraxacum officinale* (dandelion). The phloem contains numerous laticifers associated with the strands of conducting elements (1, cf., **3.35**). Copious latex is contained within the plant body and another species of *Taraxacum* was cultivated as an alternative source of rubber to the rubber tree in the Second World War. Phloem parenchyma (2), secondary xylem (3), vascular cambium (4). (LM x 100.)

1	Phloem conducting elements
2	Phloem parenchyma
3	Secondary xylem
4	Vascular cambium

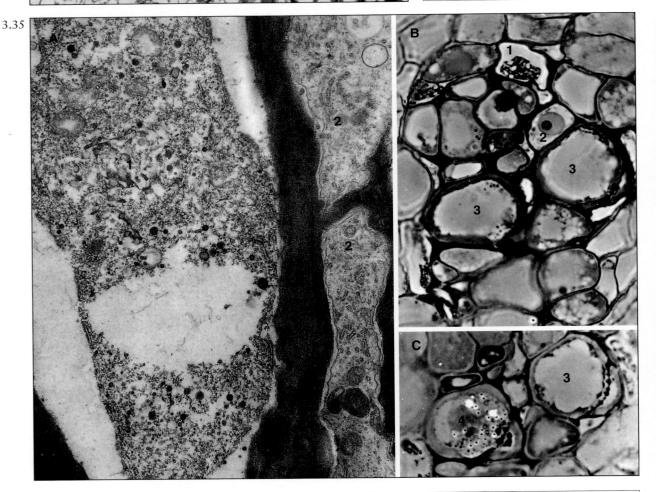

3.35 Laticifers in the root of the dicotyledon *Taraxacum officinale* (dandelion). **A**, detail of the latex in a laticifer (1) and a pair of companion cells (2). **B**, **C** show adjacent sections of the same conducting strand in transverse view of the secondary phloem. Two sieve tubes (3) are evident and in one a sieve plate is sectioned (4); these conducting elements are associated with laticifers (1) and companion cells (2). (**A**, G-Os, TEM x 12,600; **B**, **C**, G-Os, Phase LM x 1,500.)

1	Laticifer
2	Companion cells
3	Sieve tubes
4	Sieve plate

3.36 TS of the young stem of *Nerium oleander*. Note the laticifers (arrows) which permeate the pith (1) and cortex (2) and also occur in the xylem (3). (LM x 75.)

1	Pith
2	Cortex
3	Xylem

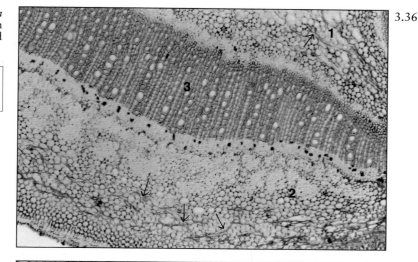

3.37 TS of a young twig of the dicotyledon *Tilia cordata* (lime) showing the secondary phloem. This is a complex tissue with wide flares of ray parenchyma cells (1) which divide tangentially to accommodate the increasing circumference of the stem as secondary thickening progresses. The conductive phloem elements (2) function over several seasons and they are interspersed with tangential bands of thick-walled fibres (3). Vascular cambium (4), secondary xylem (5). (LM x 70.)

1	Ray parenchyma
2	Phloem elements
3	Fibres
4	Vascular cambium
5	Secondary xylem

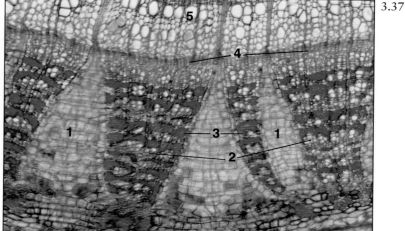

3.38 TS of the dicotyledon *Tilia cordata* (lime) stem showing detail of the phloem (cf., **3.37**). The wide sieve tubes (1) (cf., **3.37**) are sandwiched between tangential bands of very thick-walled fibres (2). (LM x 285.)

1	Sieve tubes
2	Fibres

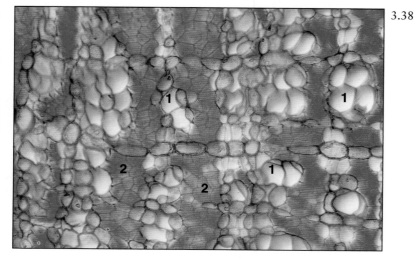

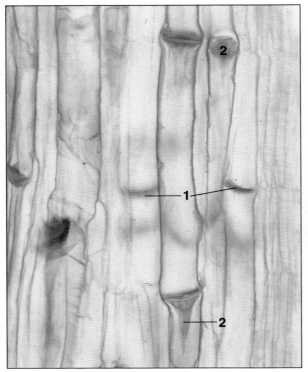

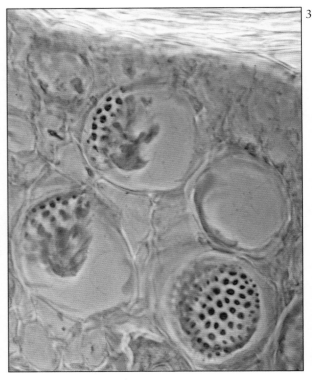

3.39 LS of the phloem of the dicotyledon *Cucurbita* showing numerous large-diametered sieve tubes in the stem. These are interrupted periodically by transverse or slightly oblique sieve plates (1). The P-protein fibrils coagulated when the specimen was excised and now form plugs (2) at the sieve plates. (LM x 335.)

3.40 TS of the phloem of the dicotyledon *Cucurbita* showing several broad sieve plates with wide sieve pores which are bordered by deposits of callose. (Phase LM x 545.)

1	Sieve plates	2	P-protein plugs

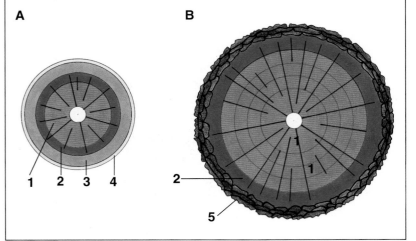

3.41 Diagrammatic representation of bark formation in a woody dicotyledonous stem. **A**, secondary xylem (1) and phloem (2) have formed but the cortex (3) and epidermis (4) are still intact. **B**, the epidermis has been replaced by a thick layer of bark (5) which protects the adjacent cylinder of secondary phloem (2).

The first-formed cork cambium generally arises hypodermally but is short lived; successive cambia arise internally from any remaining cortex and later the secondary phloem. These cork cambia are usually discontinuous and form overlapping concave shells. The outermost, non-functional secondary phloem becomes sloughed off in the various layers of the peeling bark.

1	Secondary xylem	4	Epidermis
2	Secondary phloem	5	Bark
3	Cortex		

3.42 Diagrams of a sieve element with compound sieve plates. **A**, LS showing oblique sieve plates (1) on the end walls and sieve areas (2, seen in face view) on the side walls; a nucleated companion cell (3) is also evident. **B**, detailed view of the compound sieve plate shown in **A**. Note the beaded appearance of the compound plate; the sieve element nucleus has degenerated and the few remaining organelles are greatly modified, but the plasmalemma (4) remains intact. **C**, detailed view of a compound sieve plate seen in face view.

1	Oblique sieve plates
2	Sieve areas (face view)
3	Companion cell
4	Plasmalemma

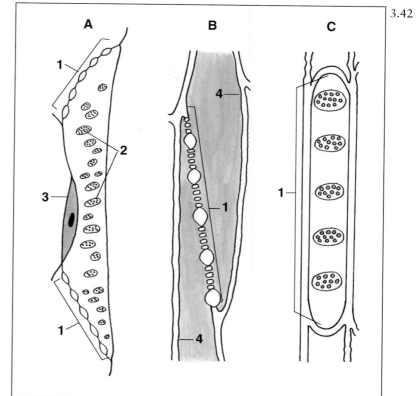

3.43 Diagram showing the pressure-flow model of translocation through the phloem from the source to the sink. Sugars are photosynthesised in the chloroplasts of the mesophyll (l) and transported in solution (both apoplastically via the walls, and symplastically through the protoplasts) until reaching a veinlet.

The solutes pass across the bundle sheath parenchyma (2) and nucleated companion cells (3) and are then actively loaded into the sieve tubes (4). The osmotic pressure of the latter increases so that water enters the system and a bulk flow of water and solutes (wide arrows) occurs towards the sink. Here active unloading takes place so that a lower turgor pressure occurs in the sieve tubes, and the sugars move in solution into the storage tissue (5) or growing regions of the root and shoot.

1	Photosynthetic mesophyll
2	Parenchyma
3	Companion cells
4	Sieve tubes
5	Storage tissue

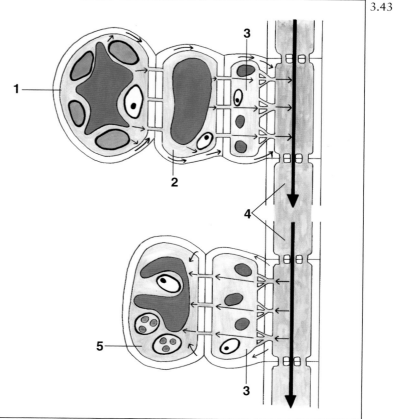

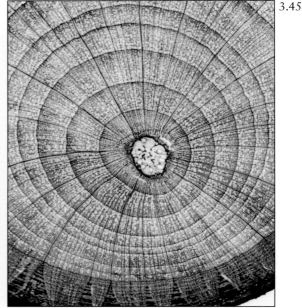

3.45 TS of a young branch of the dicotyledon *Tilia cordata* (lime tree). This shows eight clearly defined growth rings but the twig is unevenly thickened due to the formation of tensionwood on its upper side.

3.44 TS of a lateral branch of the dicotyledon *Laburnum anagyroides*. The great bulk of the branch consists of secondary xylem and about 50 growth rings are present; the non-conducting heartwood (1) is darker than the surrounding sapwood (2). On the outside a narrow, darker layer (3) of bark and secondary phloem is evident.

Numerous radial rays transverse the wood and the branch has split down one of these as the wood dried out. Note the eccentric appearance of the wood; in life the narrower (split) side of the branch corresponded to the lower surface of the branch since the production of reaction (tension) wood thickened the upper side.

1	Non-conducting heartwood
2	Sapwood
3	Bark and secondary phloem

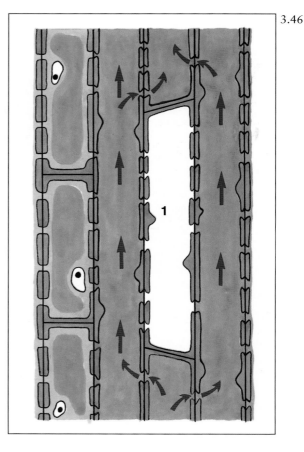

3.46 Diagrammatic representation of cavitation in the xylem of an angiosperm. The air bubble (1) has blocked movement along a vessel (which is shown as consisting of two members but could be much longer) but the embolism cannot spread. The interrupted flow of water is diverted laterally, via the unlignified pits in the tracheary walls, to adjacent vessels.

1	Air bubble

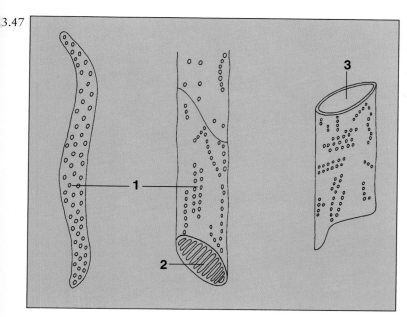

3.47 Diagrammatic representation of isolated tracheary elements as seen in macerated wood. Numerous pits (1) are present on their longitudinal walls and these are generally bordered where they interconnect with other tracheary elements. **A**, tracheid consisting of a single (thick-walled) elongate cell with tapering ends. **B**, part of a pitted vessel with an oblique, scalariform perforation plate (2) visible in face view. This vessel element is joined to another but the intervening perforation plate shows as an oblique line. **C**, isolated vessel element showing a simple, approximately transverse, perforation plate (3) at one end.

1	Pits
2	Scalariform perforation plate
3	Simple perforation plate

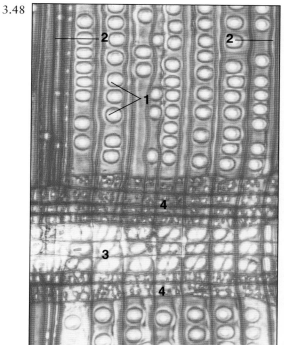

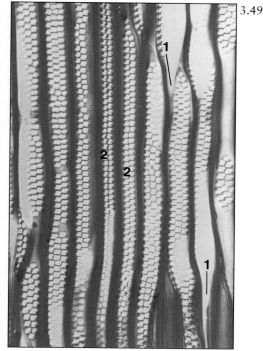

3.48 RLS of the wood in the conifer *Pinus ponderosa* (pine) stem. Note the numerous large bordered pits (1) occurring in a single row, along the radial walls of the early formed tracheids. To either side a band of narrow later wood tracheids (2) is evident. A tall ray traverses the wood at right angles to the long axes of the tracheids and both ray parenchyma (3) and ray tracheids (4) are visible. (LM. x 170.)

1	Pits
2	Late wood tracheids
3	Ray parenchyma
4	Ray tracheids

3.49 RLS of the secondary xylem of the conifer *Araucaria angustifolia*. The elongate tracheids have tapered ends (1) which interdigitate but are not perforated; however, water moves from one element to another via the numerous bordered pits (2) which occur in an alternate arrangement in their radial walls. (LM x 100.)

1	Tracheid tapered ends
2	Bordered pits

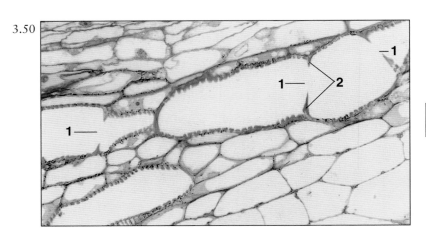

3.50 Scalariformly thickened vessels in the primary xylem of the dicotyledon *Phaseolus vulgaris* (bean). Note the several simple perforation plates (1) with the rims of the original transverse wall still evident (2). (G-Os, LM x 305.)

1	Simple perforation plates
2	Wall rims

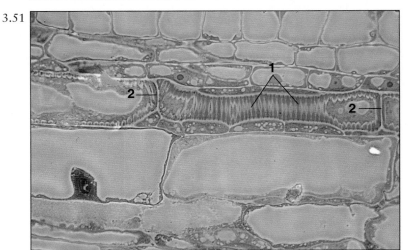

3.51 Stem of the dicotyledon *Phaseolus vulgaris* (bean) showing differentiating metaxylem elements. The narrow elements (which still retain their cytoplasm, cf., **2.8**) show scalariform wall thickening (1). Their transverse walls (2) later break down to form simple perforations (cf., **3.50**). (G-Os, Phase LM x 305.)

1	Scalariform wall thickening
2	Transverse walls

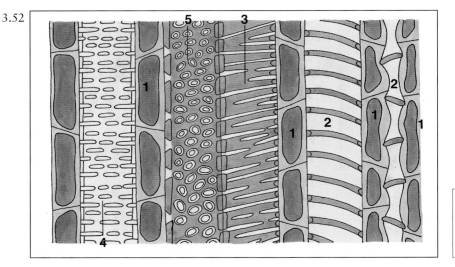

1	Axial parenchyma cells
2	Annular protoxylem
3	Scalariform metaxylem
4	Reticulate metaxylem
5	Pitted metaxylem

3.52 Diagrammatic LS through the primary xylem of a dicotyledon. Xylem is a complex tissue (cf., **3.5**) and living axial parenchyma cells (1) are interspersed with dead tracheary elements. The latter consists of: I. protoxylem in which lignified secondary wall is deposited internal to the cellulosic primary wall in an annular (2) or spiral pattern; II. metaxylem in which the lignified secondary wall is much more extensive and is deposited in a scalariform (3), reticulate (4) or pitted (5) pattern.

3.53 TS of the wood of the conifer *Thuja* showing radially-aligned tracheids. The wide lumina of the early tracheids (1) contrast with their narrow lumina in the late wood (2). Note the narrow ray (3) traversing the wood, and the pits (4) on the radial walls of the tracheids. (LM x 275.)

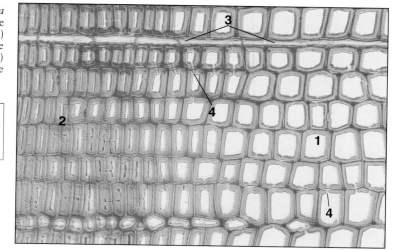

1	Early tracheids
2	Late wood
3	Ray
4	Pits

3.53

3.54

3.55

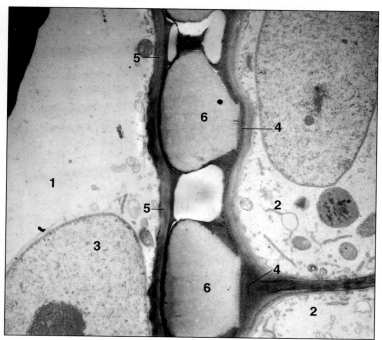

3.54 TS of the secondary xylem of the dicotyledon *Aucomea klaineana*. This is native to West Africa and, as in many tropical trees, growth rings are not obvious since the large vessels (1) are evenly distributed throughout the wood. (LM x 25.)

1	Large-diametered vessels

3.55 LS through the secondary xylem on the root of the dicotyledon *Taraxacum officinale*. The lumen of the vessel is occluded by a tylosis (1) which has grown through a pit in the vessel wall from the adjacent xylem parenchyma cells (2). Nucleus of tylosis (3), primary wall vessel (4), primary wall of tylosis (5), secondary wall of vessel (6). (G-Os, TEM x 7,000.)

1	Tylosis	4	Primary wall vessel
2	Xylem parenchyma cells	5	Primary wall of tylosis
3	Tylosis nucleus	6	Secondary wall of vessel

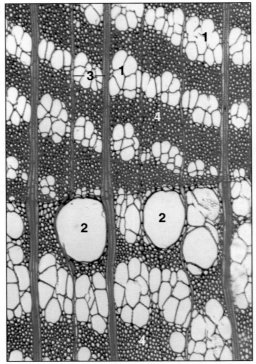

3.56 TS of the secondary xylem of the dicotyledon *Ulmus americana* (American elm). This is a ring porous wood and clearly shows the abrupt transition between the smaller vessels (1) of the previous year's later growth and the large vessels (2) formed at the beginning of the new season. Ray parenchyma (3), fibres (4). (LM x 75.)

1	Smaller last season's vessels
2	Larger new season's vessels
3	Ray parenchyma
4	Fibres

3.57 TLS of the secondary xylem of the dicotyledon *Ochroma lagopus* (balsa). The abundant and large parenchymatous rays (1), the thin-walled fibres (2), and the relatively few and thin-walled vessels (3) make this wood very light (specific gravity 0.1–0.16). Perforation plate (4), alternately arranged pits in vessel (5). (LM x 75.)

1	Parenchymatous rays
2	Thin-walled fibres
3	Thin-walled vessels
4	Perforation plate
5	Pits

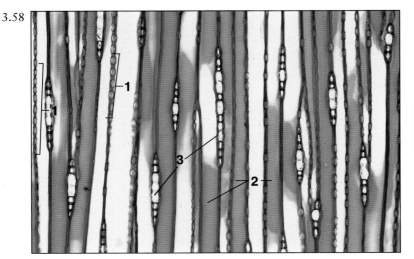

3.58 TLS of the conifer wood of *Pinus* (pine). Note that the bordered pits (1) occur in the radial walls of the tracheids but are absent from the tangential walls (2). The numerous non-storied rays (3) are only a single cell wide but extend a number of cells in height (cf., **3.48**). (LM x 110.)

1	Bordered pits in radial walls
2	Tangential walls (no pits)
3	Non-storied rays

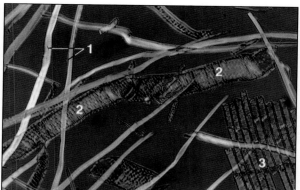

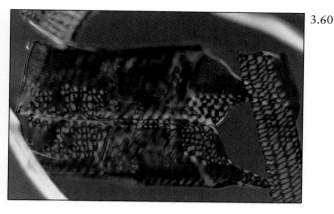

3.59 Macerated wood from the dicotyledon *Quercus alba* (oak). Note the long, narrow fibres (1) with few pits; these contrast with the wide-diametered vessel elements (2) showing numerous pits. Ray parenchyma cells (3) are also evident. (Polarised LM x 95.)

3.60 Macerated wood from the dicotyledon *Fraxinus americana* (ash). Note the disassociated vessel elements with abundant pitting on the lateral walls and the simple perforations which terminate the vessel elements. (Polarised LM x 235.)

1 Fibres	3	Ray parenchyma cells
2 Vessel elements		

3.61 TS of the wood of the ringporous stem of the dicotyledon *Fraxinus americana* (ash). Note the very wide vessels (1) in the early wood and the few narrow-diametered, single or aggregated (often paired) vessels (2) in the later wood. The latter lie in a ground mass of narrow, thick-walled fibres (3), but the fibre-tracheids (4) of the early wood have thinner walls. Narrow rays (5) traverse the wood. (LM x 110.)

1	Wide vessels
2	Narrow vessels
3	Thick-walled fibres
4	Fibre tracheids
5	Rays

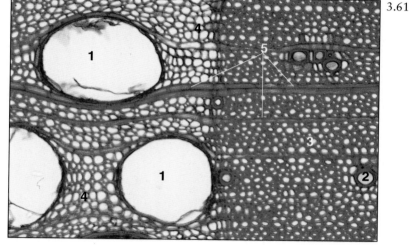

3.62 TS of the secondary xylem of the dicotyledon *Pterygota kamerumensis*. The wood shows broad transverse bands of parenchyma (1) and wide and tall multi-seriate rays (2). Vessel (3). (LM x 75.)

1	Parenchyma
2	Multiseriate rays
3	Vessel

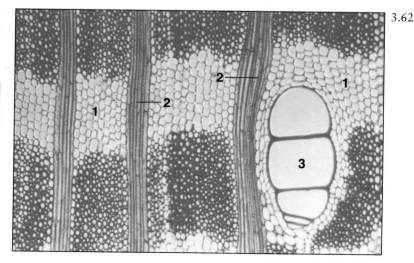

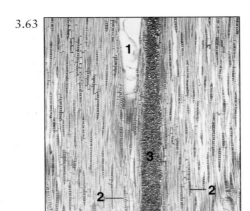

3.63 TLS of the secondary xylem of the dicotyledon *Quercus alba* (oak). Note the several wide-diametered vessels (1) containing tyloses and the numerous short uniseriate rays (2). These contrast greatly with the very wide multiseriate rays (3). Axial parenchyma bands are also present amongst the narrow-diametered tracheary elements and fibres. (LM x 25.)

1	Vessels
2	Uniseriate rays
3	Wide multiseriate ray

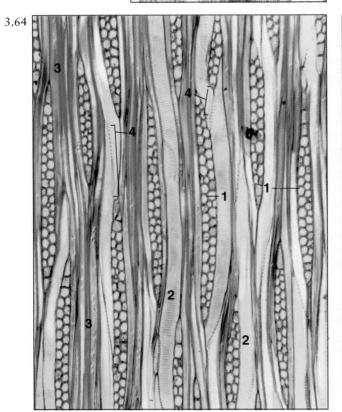

3.64 TLS of the secondary xylem of the dicotyledon *Magnolia grandiflora* (cf., **3.14**). Abundant multiseriate rays (1) occur between the uniformly diametered vessels (2) and narrower fibres (3). Scalariform perforation plate (4). (LM x 85.)

1	Multiseriate rays	3	Fibres
2	Vessels	4	Scalariform perforation plate

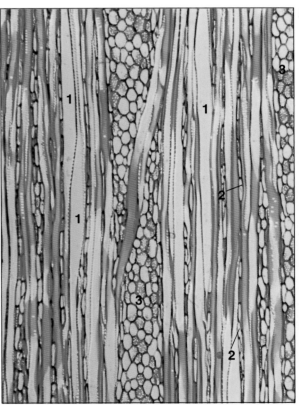

3.65 TLS of the secondary xylem of the dicotyledon *Drimys winteri*. This primitive angiosperm does not develop vessels (cf., **3.66**) but abundant tracheids (1) occur in the xylem. Both uniseriate (2) and multiseriate (3) rays are present in the wood. (LM x 85.)

1	Tracheids	3	Multiseriate rays
2	Uniseriate rays		

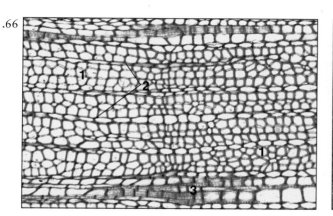

3.66 TS of the wood of the primitive dicotyledon *Drimys winteri*. This does not contain vessels and the tracheids (1) are of rather uniform diameter. Both uniseriate (2) and multiseriate rays (3) are present. (LM x 90.)

1 Tracheids	3 Multiseriate ray
2 Uniseriate ray	

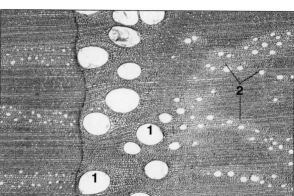

3.67 TS of the ring porous wood of the dicotyledon *Quercus alba* (oak). The large-diametered early-wood vessels (1) contrast with the smaller-diametered vessels of the late wood. The latter are distributed in more-or-less radial bands of vessels (2) alternating with areas of very narrow tracheary elements. Numerous rays are also evident. (LM x 25.)

1 Early-wood vessels
2 Late wood vessels

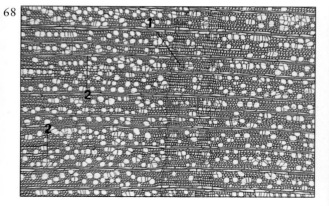

3.68 TS of the diffuse porous wood of the dicotyledon *Magnolia grandiflora*. The small and fairly uniform vessels occur evenly throughout the growth ring except for a narrow band of late wood (1). Numerous, narrow multiseriate rays are present (2). (LM x 25.)

1 Late wood vessels	2 Multiseriate rays

3.69 Trunk of a large specimen of *Cedrus deodora* (cedar) showing the scar of a large side branch. In gymnosperms (in contrast to most angiosperms, cf., 3.37) the reaction (compression) wood, forming the thicker part of the eccentric xylem, lies on the underside of the branch (1, indicates the position originally occupied by the pith). Note the callusing at the margins of the scar which is beginning to cover the wound with cork (2)

1 Original position of pith	2 Wound cork

CHAPTER 4

Apical meristems: genesis of primary shoot and root

Introduction

The germinating seedling originates from a bipolar embryo in which the radicle is situated at one pole and the plumule at the other (**1.33, 4.1**). The plumule is flanked by one or more cotyledons (**1.21, 1.33**). The meristematic cells of the root and shoot tips (**2.1, 4.1 to 4.5**) are the source of the cells which differentiate within the apical and sub-apical regions (**4.3 to 4.8**) into the primary tissues of the stem and root. The leaves and buds originate from the margins of the shoot apex (**4.3, 4.4, 4.6**) but lateral roots (**4.9**) develop some distance from the root tip. The rates of cell division, differentiation and maturation vary greatly within different regions of the root and shoot apices. Meristematic activity usually extends for some distance behind the apex.

Vegetative shoot apex

The shoot apex is usually domed or conical and is typically invested with leaf primordia (**4.6, 4.10 to 4.13**). In median longitudinal section of the angiosperm apex there is an outer tunica of one to several layers of cells. This surrounds a corpus of less regularly-arranged cells (**4.2, 4.3, 4.14**). In the tunica the cells divide anticlinally so that the newly-formed division walls always lie perpendicular to the outer surface of the apex. Thus the tunica remains discrete and its outermost cell layer differentiates into the epidermal covering of the leaves, buds and young stem (**4.3, 4.14**). By contrast, in gymnosperms the multicellular apex lacks a distinct tunica layer. The shoot epidermis is covered by cuticle and even at the apex a thin cuticle occurs (**4.2**).

In addition to zonation based upon planes of cell division (tunica-corpus), the cells within the shoot apices of some species appear cytologically heterogeneous with a core of somewhat vacuolated rib meristem cells enclosed by the more densely-staining flank meristem cells (**4.3**). In some cases the cells at the extreme tip of the apex stain less densely (but is not evident in **4.3, 4.14**) and have been designated central mother cells. These apparently divide much more slowly than those of the flank and rib meristems. The rib meristem consists of vertical rows of vacuolating cells (**4.3**) which differentiate into the central ground tissue (pith) in the young stem of dicotyledons (**4.6, 4.7**), while the flank meristem (**4.3**) gives rise to the epidermis, cortex and procambium (**4.4, 4.6, 4.7**). However, no rigid boundary occurs between the rib and flank meristems so that the cortical and procambial tissues are somewhat variable in their derivation from the apex.

Early leaf and bud development

Leaves are initiated from the margins of the shoot apex (**4.14**) by anticlinal divisions in the outermost tunica layer and variously oriented divisions in the internal tissue. The arrangement of leaves relative to the shoot apex and stem (phyllotaxy) is usually characteristic of the species (**4.11, 4.13**), and sometimes of the family, as in grasses and labiates (**1.23, 4.6, 4.10**). In monocotyledons the leaf primordia are initiated singly from the shoot apex and the leaf base extends around a broad arc of the apex (**1.23, 4.15**). By contrast, in dicotyledons a leaf primordium usually develops from a narrower sector of the apex and initially forms a peg-like protuberance (**4.11, 4.12, 4.16**).

In dicotyledons the leaf primordia may be initiated singly from the shoot apex giving rise to a spiral phyllotaxy (**4.11, 4.13**), in opposite pairs (**4.3, 4.10**) or in whorls of three or more primordia (**4.12**). In grasses, and commonly in other monocotyledons, the leaf arrangement is distichous (**1.23**) and they lie in two rows which are usually 180 degrees apart. However, spiral and other leaf arrangements also occur in monocotyledons.

Axillary bud primordia are frequently delimited as lateral meristematic swellings close to the shoot apex, at the adaxial junction of the young leaf and stem (**4.3, 4.10**), and are linked by procambium to the vascular supply of the young shoot (**4.17**). However, dichotomous branching, which results from division of the shoot apex into two equal parts, occurs in several angiosperm families and is common in many non-seed-bearing vascular plants. In some angiosperms axillary buds may be absent or may develop some distance behind the shoot apex from axillary parenchyma. In many species one or more accessory buds develop in addition to the axillary buds (**4.18**).

Buds may also develop adventitiously on the shoot: when the plumule of *Linum* is damaged the hypocotyl forms adventitious replacement buds (**4.19**) from dedifferentiated epidermal and cortical tissue (**4.20**). Adventitious buds are common on other stems and may also develop on roots and leaves (**3.19** to **3.21**). Shoots frequently develop from dormant buds located on the trunk or main branches of trees (**4.21**); these buds are commonly adventitious and arise endogenously from vascular parenchyma or cambial tissue. A number of tropical trees (e.g. *Artocarpus, Theobroma*) are cauliflorous, developing their flowers from persistent bud complexes on the mature trunk.

In many tropical species, the axillary buds develop into lateral shoots just beneath the terminal bud (sylleptic growth). However, in other plants the terminal bud exerts dominance over the axillaries; these are commonly invested by bud scales and undergo a period of dormancy before sprouting (proleptic growth, **4.22**).

Tissue differentiation in the young stem

In the terminal bud the procambium (incipient vascular tissue) develops acropetally into the apex (**4.4**) from the older procambial tissue at its base. Within the apex the procambium becomes differentiated from the inner flank meristem, with each leaf linked from its inception to the procambium (**4.3**, **4.4**). In the procambial strand the first vascular tissues begin to differentiate close to the apex (**2.3**, **4.7**). The protoxylem normally develops at the inner margin of the strand (endarchly) while protophloem forms exarchly on the margin nearest to the epidermis (**1.11**).

Protoxylem usually first differentiates within the procambial strand at the base of a leaf primordium (**4.15**), forming a short longitudinal file of tracheary elements which then differentiates bidirectionally: both upwards into the leaf and downwards into the young internode where it links with older and larger xylem strands (**4.23**). The longitudinal pathway of protophloem differentiation in the procambium is normally acropetal into the young axis and leaf primordia and is in continuity with the phloem elements in the older bud. The metaphloem differentiates somewhat later and is located inwards (centripetally) to the protophloem, while the metaxylem develops centrifugally to the protoxylem (**4.24**). In dicotyledons and gymnosperms a narrow strip of procambium remains undifferentiated between the xylem and phloem and constitutes the fasicular vascular cambium (**3.5**, **4.24**), but in monocotyledons this is absent (**1.11**).

In most dicotyledons a large parenchymatous pith occupies the centre of the primary stem and is surrounded by a ring of discrete vascular bundles, with a narrow cortex situated externally (**1.28**). In monocotyledons a distinct pith is uncommon and the vascular bundles normally occur throughout the ground tissue (**1.27**, **4.15**).

Root apex

In the great majority of species the root apex is sub-terminal since it is covered by a protective root cap (**4.25**), although in some aquatic plants this is absent. Due to massive dictyosome activity in the outer cap cells (**2.19**), a large quantity of mucigel is secreted into the soil rhizosphere (**4.26**). More mucigel is contributed by the root hairs (**4.27**) which develop behind the root apex. In some plants such as *Zea*, the root cap has its own distinct initials (calyptrogen, **4.25**). The incipient epidermis (protoderm) and cortex can be traced to a single tier of cells adjacent to the calyptrogen, while the central procambial cylinder (**4.5**) apparently originates from a third tier of initials immediately within those of the protoderm-cortex (**4.25**). However, in some taxa the protoderm and root cap originate from a common tier of initials, whilst in others the cap and other regions all converge to a common group of initials (**4.28**).

In actively growing and elongating root tips the tip of the apex represents a quiescent centre (**4.29**), so that the patterns of apical initials described previously reflect the architecture of the apex before active growth of the root primordium had commenced. In *Zea* the 'initials' of the protoderm, cortex and procambium (**4.25**) all lie within an extensive quiescent centre whose cells divide on average once every 174 hours in contrast to every 12 hours in the calyptrogen. On the surface of the quiescent centre, remote from the calyptrogen, the cells also divide rapidly and are the real initials of the protoderm, cortex and procambium. If the dividing cells of the root tip are damaged by ionising radiation, the resistant quiescent centre cells become reactivated and regenerate a new apex.

The root apex does not normally give rise to the lateral roots; instead these normally develop from the maturing root several millimetres from its tip (**4.9**), basal to the root hair zone (**4.27**).

Tissue differentiation in the young root

The longitudinal differentiation of both xylem and phloem within the procambium is acropetal.

These vascular tissues occur nearer to the apex in slow growing or dormant roots than in actively elongating roots. The first mature protophloem elements appear at the circumference of the procambial cylinder and differentiate closer to the root apex than the first mature protoxylem (**4.8**). By contrast the prospective metaxylem becomes demarcated, by its prominent vacuolation growth (**4.5**), nearer the apex than the protophloem. Some distance basal to the protophloem, mature proto-xylem elements develop between the protophloem files. Maturation of the metaphloem and metaxylem proceeds centripetally (**4.30**) on the radii already demarcated by the protophloem and protoxylem (**4.8**), and the vascular system of the mature primary root is formed (**4.31**). The outer layer of the procambium gives rise to the parenchymatous pericyle (**4.8, 4.31**).

In dicotyledonous roots the tissue between the xylem and phloem often forms a cambium (**4.31**), which later spreads laterally over the protoxylem poles to form a continuous meristem (**1.25**). The vascular cylinder is invested by the parenchy-matous cortex (**1.29, 4.30**) which frequently contains conspicuous intercellular air spaces. The pericycle is bounded by the single layered endodermis (**1.29, 4.31, 4.32**) whose radial and transverse walls are impregnated with ligno-suberin to form Casparian bands (**4.33**). These laterally continuous bands are impermeable, so that all water and solute movement across this layer is confined to the symplast (**4.33**). In the roots of many species, particularly monocoty-ledons, the ligno-suberisation of the endodermis later extends to all walls and additional cellulose thickening may be deposited (**4.32**). Such cells, however, still allow symplastic transport from the cortex to the stele via their plasmodesmata.

The endodermis is an important selective barrier but it allows the active transport into the vascular cylinder of certain beneficial ions (potassium, phosphate) absorbed from the soil. Calcium, however, apparently moves apoplastially and cannot enter the vascular cylinder through the endodermis, but instead passes into the vascular cylinder in the very immature root where Casparian bands have not developed. The active transport of ions into the vascular cylinder accounts for the phenomenon of root pressure which sometimes plays an accessory role to transpiration in the movement of water to the shoot.

In some roots the outer cortex differentiates as a one- to several-layered exodermis (**1.29**) but the Casparian bands in the radial walls are usually masked by deposition of suberised lamellae adjacent to the protoplast (**4.34**). A short distance behind the apex a zone of absorptive root hairs develops from the epidermis (**4.27**); but water absorption also occurs over the rest of the epidermal surface and in some species root hairs are absent. In the aerial roots of some epiphytic orchids and aroids the multilayered epidermis develops into a dead velamen (**4.34**), whose walls are thickened by bands of cellulose. This is probably an adaption to absorb water from the humid atmosphere of tropical forests.

4.1

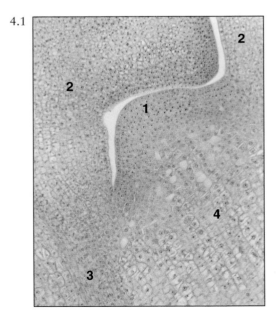

4.1 RLS of the plumule of *Phaseolus vulgaris* (bean) seed. The hemispherical, densely-staining and small-celled shoot apex (1) bears a pair of leaf primordia (2) at its margins which represent the first foliage leaves. A prominent strand of procambium (3) demarcates the longitudinal rows of vacuolated pith (4) from the cortex. (LM x 110.)

1	Shoot apex
2	Leaf primordium
3	Procambium
4	Pith

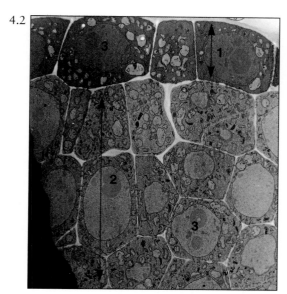

4.2

4.3

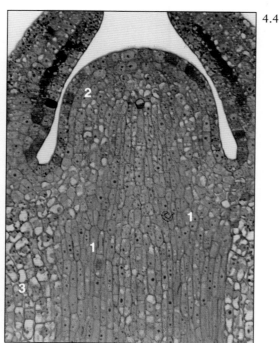

4.4

4.2 RLS of the extreme tip of the vegetative shoot apex of the dicotyledon *Glechoma hederacea* (ground ivy). Note the single tunica layer (1) where only anticlinal divisions occur and the underlying corpus (2) which divides in various planes. The small thin-walled cells possess large nuclei (3) while the cytoplasm is relatively unvacuolated (cf., **4.3**). (KMn, TEM x 1,830.)

1 Tunica layer	3 Nucleus
2 Corpus	

4.3 RLS of the vegetative shoot apex of the dicotyledon *Glechoma hederacea* (ground ivy). Note the pair of leaf primordia (1) at its base and a pair of bud primordia (2) in their axils. The extreme tip of the apex (cf., the corresponding area in **4.2**) consists of small densely cytoplasmic cells, but the rib meristem cells (3) below are vacuolating and become the pith (4) of the young stem. The margins of the apex consist of densely-staining flank meristem cells (5). Procambium (6). (G-Os, LM x 285.)

1 Leaf primordium	4 Pith
2 Bud primordium	5 Flank meristem
3 Rib meristem	6 Procambium

4.4 LS of the vegetative shoot apex of the dicotyledon *Glechoma hederacea* (ground ivy). This section is slightly tangential to the same apex shown previously (cf., **4.3**). Note how the narrow, densely cytoplasmic procambial cells (1) merge with the flank meristem (2) within the shoot apex. Cortex (3). (G-Os, LM x 285.)

1 Procambium	3 Cortex
2 Flank meristem	

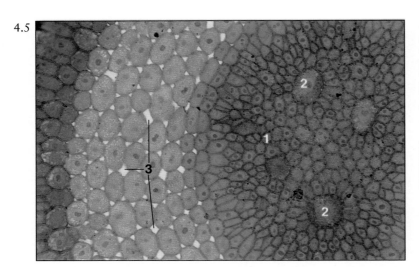

4.5 TS just behind the root apex of the monocotyledon *Zea mays* (maize). Note the core of small-diametered, densely cytoplasmic procambial cells (1) in which several larger cells (2) are blocking out large metaxylem elements. The procambium is bounded by a wide cortex of larger cells, with intercellular spaces (3) already evident. (G-Os, LM x 310.)

1	Procambial cells
2	Prospective metaxylem
3	Intercellular spaces

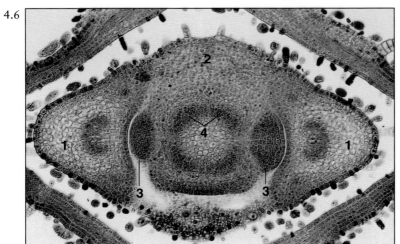

4.6 TS of the terminal bud of the dicotyledon *Glechoma hederacea* (ground ivy). The petioles (1) of the youngest pair of leaves extend as a parenchymatous collar (2) around the stem. The densely-staining cells of the axillary bud primordia (3) and procambium (4) are also evident in the stem. Note the abundant trichomes which cover the surfaces of the leaves. (LM x 125.)

1	Petioles
2	Parenchymous collar
3	Axillary bud primordium
4	Procambium

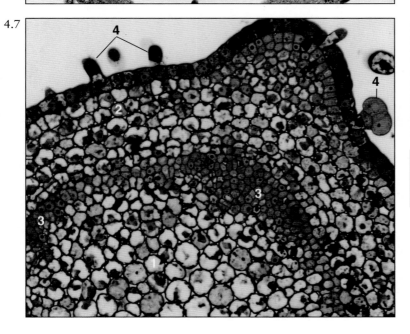

4.7 TS of the young stem of the dicotyledon *Glechoma hederacea* (ground ivy). Note the large-celled, vacuolated pith (1) and cortex (2); between them, smaller, densely cytoplasmic procambial cells (3) are evident (cf., 2.3). The epidermis is derived by anticlinal divisions from the single-layered tunica of the apex (cf., **4.2**), except that the glandular trichomes (4) form by periclinal divisions. (G-Os, LM x 370.)

1	Pith
2	Cortex
3	Procambial cells
4	Glandular trichomes

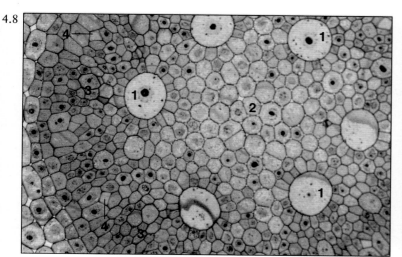

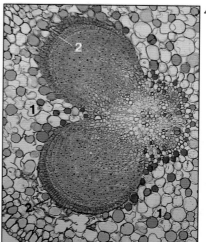

4.8 TS near the apex of the young root of the monocotyledon of *Zea mays* (maize). A number of wide-diametered, but thin-walled, potential metaxylem elements (1, cf., **4.5**) lie on the periphery of the pith (2). The sites of the potential protoxylem (3) are also evident centrifugally, but the first mature vascular elements to differentiate are the protophloem sieve tubes (4). (G-Os, LM x 310.)

1	Prospective metaxylem	3	Potential protoxylem
2	Pith	4	Protophloem sieve tubes

4.9 TS of the root of the dicotyledon *Salix* (willow). Note the presence in the parent root cortex (1) of a pair of lateral roots covered by root caps (2). These arise from the dedifferentiation and division of pericyclic parenchyma cells at the tips of the pentarch xylem arms in the parent root. (LM x 200.)

1	Parent root cortex
2	Root cap

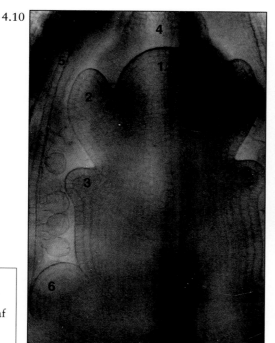

1	Shoot apex
2	Leaf primordia
3	Leaf collar
4	Face view of leaf
5	Older leaf
6	Axillary bud

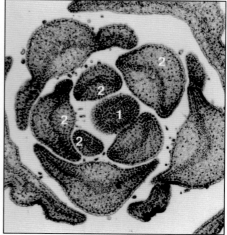

4.11 TS of the dicotyledonous vegetative bud of the dicotyledon *Solanum tuberosum* (potato). Note the shoot apex (1) and the spiral arrangement of the leaf primordia (2) which invest it. (LM x 65.)

1	Shoot apex
2	Leaf primordia

4.10 Topography of the apical region of a cleared bud of the dicotyledon *Glechoma hederacea* (ground ivy). The hemispherical shoot apex (1) shows its youngest pair of leaf primordia (2) at its margins while the collar (3, cf., **4.6**) of the next leaves is visible at the base of the apex. The tip of one of these pair of leaves is visible in face view (4). At the base of the bud an older pair of leaves (5) are seen in side view; these subtend a pair of axillary buds (6). (LM x 110.)

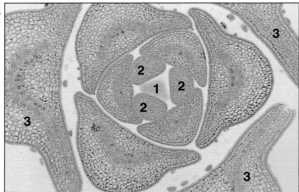

4.12

4.13

4.12 TS of the terminal bud of the dicotyledon *Ligustrum vulgare* (privet). Normally privet is decussate but in this bud a whorl of three leaves arises from the shoot apex (1). Note how the primordia of the youngest whorl of leaves (2) lie in the same relative positions as the older ones (3) arising from the axis two nodes below. (LM x 70.)

1	Shoot apex	3 Older leaves
2	Youngest leaf primordia	

4.13 Shoot tip of the succulent dicotyledon *Aeonium*. Note the closely-crowded leaves arranged in a spiral phyllotaxy (cf., **4.11**). (*Copyright of T. Norman Tait.*)

4.14

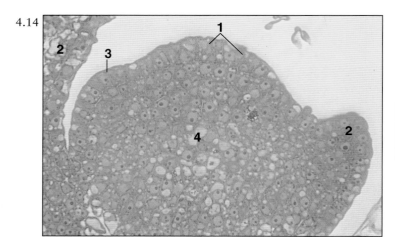

4.14 RLS of the apex of an axillary bud of the dicotyledon *Phaseolus vulgaris* (bean). The apex has a single tunica layer (1) and two leaf primordia (2) are visible while a leaf buttress (3) is also evident. Rib meristem (4) lies at the base of the apex. (G-Os, Phase contrast LM x 450.)

1	Tunica layer
2	Leaf primordia
3	Leaf buttress
4	Rib meristem

4.15

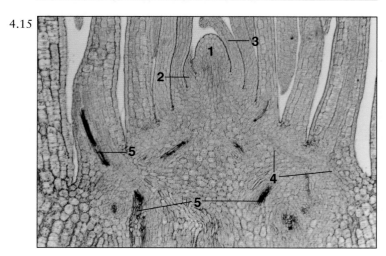

4.15 LS of the shoot apical region of the monocotyledon *Zea mays* (maize). Although at the shoot apex (1) the leaves form singly, with each new primordium (2) being initiated at 180° from the previous leaf (3), the leaf bases grow laterally and soon encircle the apex (cf., **1.23**). Note the shallow sloping sub-apical margins bearing the leaves; here the primary thickening meristem (4) divides mainly periclinally with the derivatives differentiating into parenchyma traversed by the procambial (5) strands which supply the leaves (LM x 70.)

1	Shoot apex
2	New leaf
3	Previous leaf
4	Primary thickening meristem
5	Procambial strand

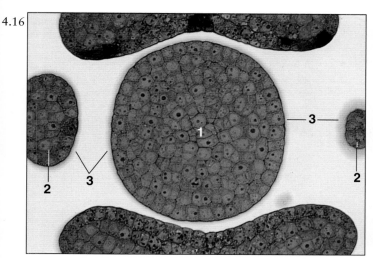
4.16

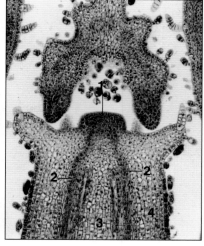
4.17

4.16 TS through the shoot apex of *Glechoma hederacea* (ground ivy) showing its decussate phyllotaxy. Shoot apex (1), youngest leaf primordia (2), cuticle (3). (G-Os, LM x 370.)

1	Shoot apex	3	Cuticle
2	Youngest leaf primorida		

4.17 TLS of the terminal bud of the dicotyledon *Glechoma hederacea* (ground ivy). An axillary bud primordium (1) is linked by densely staining procambial strands (2) to the vascular system of the internode below. Pith (3), cortex (4). (LM x 95.)

1	Bud primordium	3	Pith
2	Procambial strand	4	Cortex

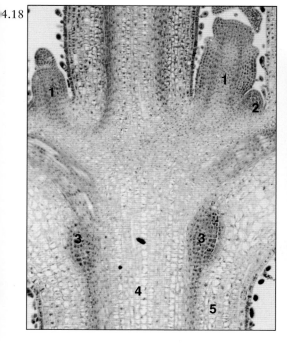
4.18

4.19

4.19 Surface view of the hypocotyl of the dicotyledon *Linum usitatissimum* (flax) showing an adventitious bud (1). Note the stomata (2) occurring in the hypocotyl epidermis. (TEM x 60.)

1	Adventitious bud
2	Stoma

4.18 LS at the base of the terminal bud of the dicotyledon *Glechoma hederacea* (ground ivy). A pair of axillary buds (1) are present, but their linkage to the vascular system of the main axis is not apparent in this plane of section (cf., **4.17**). The larger bud has an accessory bud (2) at its abaxial margin. Note the pair of adventitious root primordia (3) evident at this node. Pith (4), cortex (5). (LM x 75.)

1	Axillary bud	4	Pith
2	Accessory bud	5	Cortex
3	Root primordium		

4.20

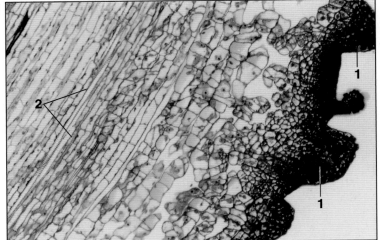

4.20 LS of the hypocotyl of the dicotyledon *Linum usitatissimum* (flax) bearing an adventitious bud primordium. Note the numerous newly-formed cells in the underlying cortex; further division gives rise to procambial strands which link the leaf primordia (1) to the vascular tissue (2) of the parent hypocotyl (G-Os, LM x 35.)

1	Leaf primordia
2	Vascular tissue of hypocotyl

4.21

4.22

4.23

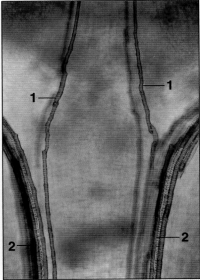

4.21 Base of the trunk of the dicotyledon *Aesculus hippocastanum* (horse chestnut). Note the numerous newly-sprouted leafy twigs; these are of adventitious origin and arise from proliferated parenchyma in the bark.

4.22 Sprouting terminal bud of the deciduous tree *Acer pseudoplatanus* (sycamore). This exerts dominance over the dormant axillary buds (1) at the nodes below. Note the decussate arrangement of the scale leaves (2), which are evanescent and non-photosynthetic and the emerging foliage leaves (3). (*Copyright of T. Norman Tait*).

1	Axillary bud
2	Scale leaves
3	Foliage leaves

4.23 Old node from a cleared terminal bud of the dicotyledon *Glechoma hederacea* (ground ivy). Note the linkage of the protoxylem (1) from the younger shoot (composed of single files of short tracheary elements) with the more extensively developed xylem strands (2) of the older stem. (LM x 100.)

1	Protoxylem files
2	Xylem strands

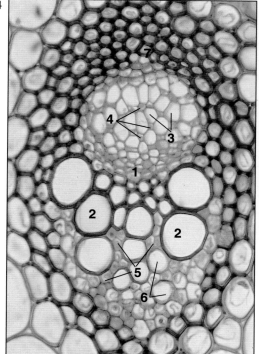

4.24

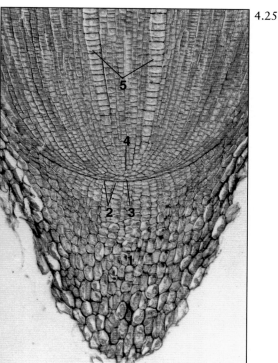

4.25

4.24 TS of a vascular bundle from the stem of *Ranunculus* (buttercup). This herbaceous dicotyledon does not undergo secondary thickening despite the presence of fascicular cambium (1) lying between the metaxylem (2) and metaphloem. The latter shows a distinctive pattern of large lumened-sieve tubes (3) and smaller companion cells (4) reminiscent of many monocotyledons (c.f., **1.11**). Axial parenchyma (5), protoxylem (6), protophloem fibres (7). (LM x 285.)

4.25 RLS of the root tip of the monocotyledon *Zea mays* (maize) showing a prominent cap covering the apex. The cap (1) has its own distinct initials (calyptrogen, 2) while the epidermis and cortex apparently arise from a common tier of initials (3) adjacent to the calyptrogen. The procambial central cylinder has its own initials (4). Note the conspicuous files (5) of enlarged cells within the procambium which represent the future metaxylem elements (cf., **4.8**). (LM x 145.)

1	Fascicular cambium	5	Axial parenchyma
2	Metaxylem	6	Protoxylem
3	Sieve tubes	7	Protophloem fibres
4	Companion cells		

1	Root cap	4	Procambial initials
2	Cap initials	5	Potential metaxylem
3	Initials of epidermis and cortex		

4.26 TS showing detail of the root cap of the monocotyledon *Zea mays* (maize). Note the progressive increase in thickness of the cell walls towards the margin of the cap. Here the cells are breaking down and sloughing their mucilaginous cell walls and protoplasts to form mucigel (1) which is secreted into the soil. (G-Os, LM x 510.)

1	Mucigel

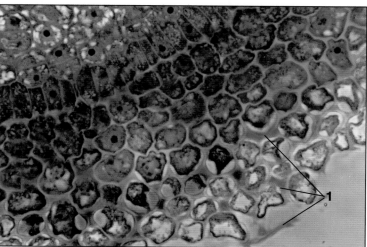

4.26

4.27 Adventitious roots on a leaf of the dicotyledon *Saintpaulia ionantha*. The excised leaf was grown *in vitro* on a non-sterile mineral salt medium and shows prolific root production at the cut base of the petiole. Note their dense felty covering of root hairs.

4.28 RLS of the radicle of the dicotyledon *Phaseolus vulgaris* (bean). This has an ill-defined group of initials which are common to the root cap (1), cortex and epidermis (2) and central cylinder (3). (G-Os, LM x 120.)

1	Root cap	3	Central cylinder
2	Epidermis and cortex		

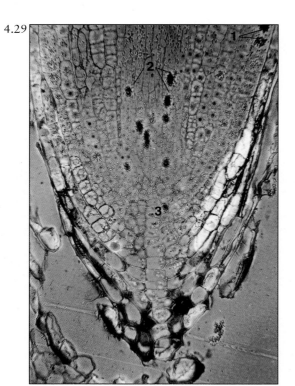

4.29 Autoradiograph of the root tip of the dicotyledon *Comptomia*. In this non-leguminous nitrogen-fixing species the root nodules sometimes elongate into normal roots. Such a root was fed with tritiated thymidine, and the subsequent autoradiograph showed heavy labelling of the nuclei in the cortex (1) and the procambium (2). Quiescent zone (3). (G-Os, Phase LM x 130.)

1	Labelled cortial nuclei
2	Labelled procambial nuclei
3	Quiescent zone

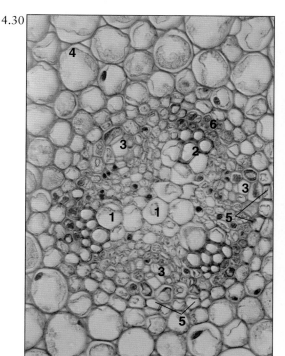

4.30 TS of the young root of the dicotyledon *Ranunculus* (buttercup). Note the triarch arrangement of its xylem, with the large, thin-walled cells in the centre representing differentiating metaxylem elements (1); at the poles the protoxylem elements (2) have already undergone secondary wall deposition and lost their protoplasts. Protophloem sieve tubes (3) have differentiated between the protoxylem poles. Cortex (4), immature endodermis (5), pericycle (6). (LM x 300.)

1	Differentiating	4	Cortex
	metaxylem	5	Immature endodermis
2	Protoxylem elements	6	Pericycle
3	Protophloem sieve tubes		

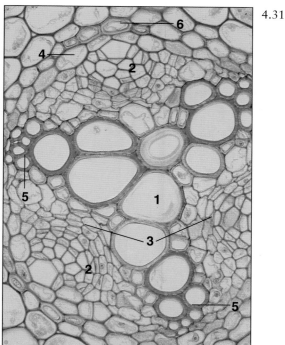

4.31 TS of the mature vascular tissue in the primary root of the dicotyledon *Ranunculus* (buttercup). Large-diametered, thick-walled metaxylem elements (1) have differentiated from the central procambium (cf., 4.30) and a number of metaphloem sieve tubes are also evident (2). The vascular cambium (3) occurs as three discontinuous arcs between the xylem and phloem. A single-layered pericycle (4) lies external to the protophloem and protoxylem poles (5). Endodermis (6). (LM x 300.)

1	Metaxylem elements	4	Pericycle
2	Metaphloem	5	Protoxylem
3	Vascular cambium	6	Endodermis

4.32 TS showing detail of the mature endodermis and vascular tissue of the root of the monocotyledon *Iris*. In the cortex (1) both apoplastic and symplastic transport of water and solutes occurs. However, movement across the endodermis (2) is symplastic through plasmodesmata in the outer tangential walls and across the protoplasts to plasmodesmata in the inner tangential walls. Pericycle (3), protoxylem (4), phloem sieve tubes (5). (LM x 305.)

1	Cortex
2	Endodermis
3	Pericycle
4	Protoxylem
5	Sieve tubes

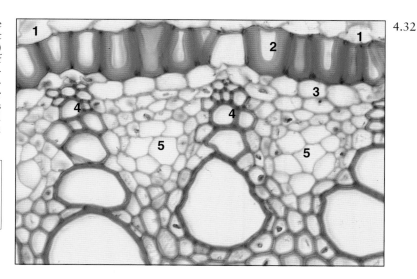

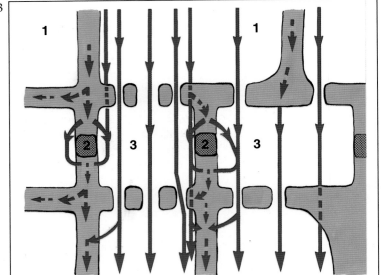

4.33 Diagrammatic representation of water and solute movement across the young endodermis. Movement from the soil through the cortex (1) is both apo- and symplastic (broken and solid blue arrows respectively). However, the impermeable Casparian bands (2) in the radial walls of the endodermis prevent apoplastic movement; water and solutes must traverse the protoplast (3) but can then move symplastically, via the plasmodesmata, or apoplastically into the vascular cylinder.

1	Cortex
2	Casparian bands
3	Endodermal protoplast

4.34

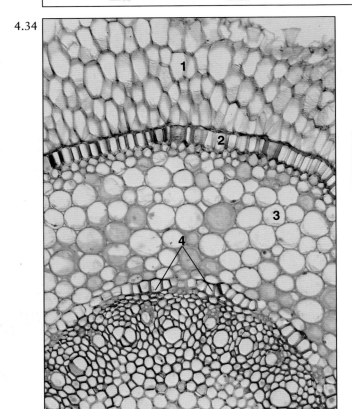

4.34 TS of the aerial root of the epiphytic orchid *Dendrobium* (a monocotyledon). A multilayered velamen (1) arises from the epidermis; its dead cells are strengthened by bands of cellulosic thickening and this tissue may absorb water from the moist tropical air. Internally a single-layered, thickened exodermis (2) is present and all water entering the cortex (3) must first move symplastically across this barrier. Endodermis (4). (LM x 80.)

1	Velamen
2	Exodermis
3	Cortex
4	Endodermis

CHAPTER 5

The green leaf

Introduction

Although the mature size and form of leaves is very variable (**1.17**, **1.19**, **3.11**, **4.22**), they all have a common origin from the margin of the shoot apex (**4.1**, **4.10**). During the life cycle of a flowering plant several types of leaf are generally developed, but only foliage leaves (**5.1** to **5.6**) are considered in detail in this Chapter. In plants showing proleptic growth (Chapter 4), scale leaves often invest and protect the immature foliage leaves within the vegetative buds (**4.22**, **5.8**) while bracts are associated with flowers and inflorescences (**5.7**). The cotyledons are leaves which become demarcated early in embryogenesis (**1.33**) and in the mature seed often contain reserve food supplies for the seedling on germination (**2.42**, **2.54**).

Morphology and venation

The foliage leaf usually has a flattened lamina or blade (**5.1** to **5.4**, **5.6**). In many dicotyledons, and some monocotyledons, the blade is borne on a petiole (**1.17**, **1.19**, **5.3**, **5.10**) whereas in most monocotyledons it is inserted directly on the stem (**5.6**). The base of the leaf sometimes encircles the stem (**1.23**, **4.6**) and a distinct leaf sheath occurs in many monocotyledons (**5.9** to **5.11**). In dicotyledons, paired stipules may occur at or near the base of the petiole (**1.19**) and these are sometimes large and vascularised. In many grasses a pair of lateral auricles and a median membranous ligule are present at the base of the lamina (**5.11**). A median hastula is present in a similar position in fan-leaved palms. A simple leaf has a single blade (**5.1** to **5.3**, **5.6**) which is often dissected or lobed (**1.17**, **5.3**). In a compound leaf (**1.19**, **5.4**, **5.10**) several to numerous leaflets occur and these may be dissected or further divided (**1.7**).

The leaf consists of dermal, ground and vascular tissues with the xylem usually orientated towards the original adaxial surface (**4.6**). In a transverse section many angiosperm leaves appear bifacial since the ground tissue of the leaf primordium (**4.11**) differentiates into the adaxial palisade and abaxial spongy mesophyll layers (**3.3**). Also, the upper and lower epidermises often differ in the thickness of their cuticle and frequency of stomata and trichomes (**5.12**). By contrast some leaves are isobilateral; here the adaxial and abaxial surfaces are similar in organisation but the xylem lies adaxially (**5.13**). In other species the leaves are unifacial with an investing epidermis of ad- or abaxial origin and usually have either a cylindrical form (**5.5**) or are sword-shaped (**5.2**, **5.6**, **5.14**). However the bases of such leaves are usually bifacial (**5.14**).

The sword-like blade of *Iris*, *Sansevieria* and similar monocotyledons is delimited by an epidermis of abaxial origin since the adaxial surfaces of the folded leaf have become confluent (**5.14**). Two series of vascular bundles are often distinguishable in thicker regions of the blade and in both series the xylem lies remote from the surface (**5.14**, **5.15**). Many monocotyledonous leaves have basal meristems from which new tissue is continuously formed; in woody perennials such as *Dracaena*, *Cordyline* and *Pandanus* (**1.22**, **5.2**), the increasing width of the leaf base is accompanied by the differentiation of new veins within the derivatives of the basal meristem.

In most monocotyledons the main veins run parallel to each other along the lamina (**1.24**, **5.16**) but eventually merge at the apex, while along the blade smaller veins, usually unbranched, connect the main veins (**1.24**). By contrast a complex reticulate venation characterises the great majority of dicotyledons and a few broad-leaved monocotyledons (**5.16** to **5.19**). Many dicotyledonous leaves possess a thickened longitudinal midrib (**1.17**) from which secondary veins branch and often extend to the laminal margins (**5.19**), while tertiary and subsequent order veins form an anastomosing minor network (**1.20**, **5.19**) in which the finest veins terminate blindly in the mesophyll (**5.17**, **5.18**).

In gymnosperms the leaf vasculature is variable; with the exception of *Cycas* (**1.5**), the cycads and their fossil ancestors show a dichotomously-branched, parallel venation in the leaflets (**5.20**). In conifers with needle-shaped leaves a single, or sometimes paired, unbranched median vein occurs

(**1.15. 5.21**). In some gymnosperms (e.g. *Cycas*, *Taxus*) files of thick-walled and elongate transfusion tracheids branch off from the main vein into the mesophyll (**5.21, 5.22**). Most ferns show dichotomous venation (**5.4**).

Anatomy of the lamina
Leaf epidermis
In angiosperms the epidermis is directly derived from the outermost tunica layer of the shoot apex (**4.3**). In a few genera (e.g. *Ficus*, *Nerium*, *Peperomia*) the immature adaxial epidermis undergoes periclinal divisions to form a multiple epidermis, but it is not possible to distinguish this from a normal hypodermis in the mature leaf (**5.12, 5.23**). The epidermis is persistent even in long-lived, evergreen leaves and it is normally coated externally by the relatively impermeable layers of cuticle and wax (**3.25, 3.26, 5.21, 5.23, 5.24**).

The epidermis always contains ordinary parenchyma-type cells together with the guard cells which surround the stomatal pores (**5.15, 5.25, 5.26**). However, stomata are often confined to the abaxial surface in bifacial leaves (**3.3, 5.21**). Additional cell types frequently occur, especially subsidiary cells associated with the guard cells (**5.27, 5.28, 5.30**) and various trichomes (**3.11, 5.29**). Chloroplasts are usually only present in the guard cells (**5.25**); whilst leucoplasts are normally present elsewhere in the epidermis (**2.47**). However, chloroplasts may differentiate in the ordinary epidermal cells of shade plants and also in aquatics.

When viewed from the leaf surface the anticlinal (vertical) walls of ordinary epidermal cells are sometimes sinuous (**5.28**). In most monocotyledonous leaves the long axes of the epidermal cells are parallel to that of the leaf (**5.28**). In dicotyledonous leaves the cells are more randomly arranged (**5.29, 5.30**), but over the midrib and main veins the epidermal cells tend to lie parallel to each other. In many xeromorphic grasses (that is species adapted to dry habitats) large, thin-walled bulliform cells occur in the adaxial epidermis (**5.31 to 5.33**). These cells preferentially lose water and so contribute to the rolling of the leaf (**5.31**). In xeromorphic plants the stomata often occur in grooves or pits (**3.22**) and epidermal trichomes are frequently present (**5.12, 5.32**). Stomata are usually absent from the epidermis where it overlies the hypodermal sclerenchyma (**3.22, 5.32**) which often occurs in extensive tracts at the margins of leaves or associated with large veins (**5.34**).

In bifacial leaves stomata are usually most frequent abaxially (**3.3, 5.21**). On the other hand in aquatics with floating leaves (**5.1**) they are usually confined to the adaxial surface, while in submerged leaves stomata are generally absent. Stomata are often randomly orientated (**5.30**) but in elongate leaves the guard cells are usually parallel (**5.28**). In grasses and sedges a pair of dumbbell-shaped guard cells surround the pore (**5.27, 5.28**), while crescent-shaped guard cells occur in most other species (**5.27, 5.30**).

In the latter the anticlinal walls remote from the pore are relatively thin but the walls adjacent to the pore are often thickened (**5.25, 5.27**). The differential thickening of these walls, and the radial distribution of cellulose microfibrils in their periclinal walls, is apparently associated with the pore opening when the guard cells are turgid. In the dumbbell-shaped guard cells of grasses the ends are thin-walled while the side walls are thicker (**5.27**). An increase of turgor in the guard cells causes their ends to swell and the anticlinal side walls are pulled apart at the pore.

A cuticle is always present on the outer periclinal wall of a guard cell (**5.25, 5,26**) and wedges of cuticle-covered wall sometimes protrude towards the pore. In *Pinus* and some other gymnosperms the guard cell and associated subsidiary cell walls are apparently lignified (**3.22**). In most species where subsidiary cells are present they do not differ from ordinary epidermal cells in their cytological appearance, but are normally distinguished by their shape and orientation relative to the guard cells (**5.27, 5.28, 5.30**). Subsidiary cells may be derived from the same precursor cell as the guard cells (mesogenous development) or from neighbouring epidermal cells (perigenous origin): the two types are not homologous and their occurence can be of systematic significance. It is not known whether subsidiary cells have a distinctive role in stomatal movement and they are apparently absent from leaves of a number of species.

Trichomes are common on leaves and stems; they range from unicellular to multicellular structures (**5.12, 5.29, 5.31**) and may be branched. Many are secretory and these often consist of a stalk and glandular head (**3.11, 4.6, 4.7**). Their secretions are diverse and vary from essential oils as in *Lavandula* to hydrolytic enzymes in the leaves of carnivorous plants such as *Drosera* (**3.11, 3.12**). In some glands the secretion may temporarily accumulate between the outer periclinal wall and the overlying cuticle. Several types of trichomes may be present on the same leaf (4.6). Non-glandular trichomes have several possible functions in the leaf, such as slowing the rate of transpiration by impeding wind movement over its surface, deterring insect attack and protection against excessive sunlight.

The outer wall of the leaf epidermis is impregnated with cutin, while the cuticle lies externally (5.21, 5.23, 5.24, 5.26). Cutin is a highly hydrophobic lipid polyester of high molecular weight. The cuticle is especially thick in xerophytes (3.25, 3.26, 5.24, 5.35). It can reach 6 mm in thickness and a thin deposit may sometimes be present on the anticlinal and innermost periclinal epidermal cell walls. The cuticle is very thin or absent in submerged shoots of aquatics. In bifacial leaves the adaxial cuticle is usually thicker than the abaxial (5.36). Wax is usually present on the surface of the cuticle where it occurs as a crust of filaments, granules, or flakes. However, the wax is often dissolved from the surface during the fixation and processing of material for examination under the microscope.

Mesophyll and sclerenchyma

In most bifacial leaves a layer of palisade cells occurs adaxially (3.3, 5.21, 5.36). These highly vacuolate, cylindrical cells are anticlinally elongated and contain numerous chloroplasts in the peripheral cytoplasm adjacent to their walls (2.24, 3.3). There is often a well-developed system of intercellular spaces allowing gaseous diffusion through the apoplast in relation to photosynthesis and transpiration. The palisade tissue may be several cells thick and in isobilateral leaves usually occurs both ad- and abaxially (5.13). Xeromorphic leaves often show a more compact mesophyll in which the intercellular spaces are reduced (3.22, 3.26, 5.31, 5.32).

In bifacial leaves a layer of spongy mesophyll occurs abaxially (3.3, 5.21). In this non-homogenous tissue, with its large intercellular spaces, the total volume of apoplast usually exceeds the symplast (5.26). However, the surface area of wall in this tissue is often much less than in the palisade mesophyll. In some xeromorphic plants and succulents the adaxial palisade is completely or partly replaced by compact non-photosynthetic parenchyma, whose large cells are highly vacuolate and probably represent a water storage tissue (5.23, 5.37). In thick leaves, the central tissue may

be achlorophyllous (5.34) and sometimes degenerates (5.15).

Differential patterns of colouring commonly occur in leaves, particularly in decorative foliage plants. Variegated, chimaeral leaves (5.6, 5.38, 5.39) usually originate from a nuclear mutation in the shoot apex which prevents derivatives of the mutated cell from developing chloroplasts. If the mutation occurs in the inner tunica or corpus, it may result in the formation of achlorophyllous tissue (5.39) in place of normal green mesophyll (3.3).

In *Pinus* and several conifers the mesophyll is plicate (3.22); vertical invaginations from the main wall protrude into the protoplast and thus increase the surface area of cytoplasm occupied by the chloroplasts. Many tropical grasses and various other taxa photosynthesise by the C4 pathway and these leaves often show a 'Kranz' (wreath) anatomy, with the mesophyll cells radiating outwards from the bundle sheaths enclosing the veins (5.40). The chloroplasts within the sheath are usually agranal and commonly larger than the granal mesophyll chloroplasts.

The margins of many leaves are strengthened by tracks of fibres and prominent strands or girders of sclerenchyma may also be interspersed in the mesophyll (3.26, 5.34). Additionally, sclereids may intrude within the mesophyll (5.35). Sclerenchyma and collenchyma are also frequently present in the ribs developed over the major veins and the mid-rib (5.36, 5.40).

Vascular tissue

In the lamina of dicotyledons the finest branches of the minor veins delimit the areoles (5.17), but in monocotyledons the minor network is less well defined (1.24, 5.16). Minor veins are embedded in a sheath of photosynthetic mesophyll (3.3) but the larger veins are often enclosed within ground tissue containing few chloroplasts (5.15, 5.36, 5.41, 5.42). The veins are typically collateral (5.41 to 5.43) with the xylem normally located adaxially (5.14B, 5.36) but their orientation may vary (5.41). Limited secondary thickening may occur in the main veins of some dicotyledonous leaves and conifer leaves (5.21, 5.36, 5.43).

The minor veins (5.17 to 5.19) are concerned with the loading of sugars formed by photosynthesis into the phloem (3.43) and the unloading of water from the xylem into the mesophyll. Vascular parenchyma and transfer cells (2.63) are especially well developed in relation to the phloem and some represent companion cells. The minor veins are enclosed within compact bundle sheaths (**2.9**, **3.3**, **5.18**) so that the vascular tissue is not normally directly exposed to the intercellular space system of the mesophyll. The sheath is often composed of parenchyma but sometimes a several-layered sheath is formed from parenchyma and sclerenchyma. The minor veins usually contain protoxylem tracheids (**5.17**, **5.18**) but not vessels, while sieve elements are often lacking in the finest veins.

Anatomy of the petiole

In many leaves the vascular supply from the stem passes to the leaf blade via the petiole (**1.17**, **1.19**, **5.3**, **5.10**). The simplest vascular arrangement is a single, large crescent-shaped vein with adaxial xylem but several veins are often present (**5.44**). The peripheral ground tissue commonly contains collenchyma and sclerenchyma (**3.23**, **5.44**). The leaves of many plants lie extended in day-light (**5.45**) but undergo a slow drooping at night (**5.46**) in response to various external or internal stimuli. This movement results from loss of turgor in the extensive zone of peripheral parenchyma located in a joint-like thickening (pulvinus) present at the base of the petiole, or at the bases of petiolules in compound leaves. The more rapid response of the leaf of the sensitive Mimosa (*M. pudica*) to touch is also due to the activity of the numerous pulvini.

In both deciduous and evergreen species an abscission zone occurs at the base of the petiole (**5.47**). Here the xylem is often restricted to tracheids or narrow, often discontinuous, vessels (**5.48**). Such tracheary elements reduce the risk of embolism when the blade is wind damaged or eaten and also when the senescent leaf abscises. At the basal face of the abscission zone a suberised protective layer of cork develops which covers the scar (**5.49**) formed when the petiole becomes detached. In some perennial monocotyledons several abscission zones occur and in palms the leaf base may persist after the petiole and lamina are detached (**5.50**).

5.1
5.1 Large floating leaves of the water lilies *Nymphaea gigantea* (foreground) and *Victoria amazonica* (rear) which are connected by long petioles to the rootstock growing in the mud on the bottom of the pond. The leaf stomata are confined to the adaxial surface which is exposed to the atmosphere and the spongy mesophyll on the abaxial leaf surface has very extensive air spaces to assist with aeration, whilst large branched sclereids strengthen this tissue. In *Victoria* large spines occur on the ribs of the abaxial epidermis and the leaves are capable of bearing the weight of a moderate-sized adult human.

5.2
5.2 Clusters of sword-shaped leaves borne on the evergreen monocotyledon *Pandanus*. Due to a diffuse secondary growth this species can grow into large trees (cf., **1.26**).

5.3 Large fan-shaped of the monocotyledon *Licuala grandis*. This palm is indigenous to S. E. Asia and Australia and bears simple leaves on long petioles.

5.4 Part of the compound leaf of the fern *Angiopteris*. Note the longitudinal midribs in each pinna from which lateral, dichotomously-branched veins supply the lamina.

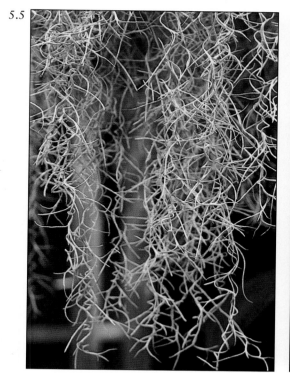

5.5 Hanging mass of the tropical epiphytic flowering plant *Tillandsia usneoides* (Spanish moss). This is rootless but the numerous fine, cylindrical leaves absorb water from the moist atmosphere.

5.6 Variegated plant of *Chlorophytum comosum* (spider plant). The elongated, sword-shaped leaves are typical of a monocotyledon. Note the leafy plantlet (1) which has developed adventitiously on the inflorescence stalk.

1 Leafy plantlet

5.7

5.7 Inflorescence of the dicotyledon *Digitalis purpurea* (foxglove). Each flower is subtended by a small green bract while five green sepals lie at the flower's base. The petals are united into purple bells whose spots probably act as honey-guides for visiting insects.

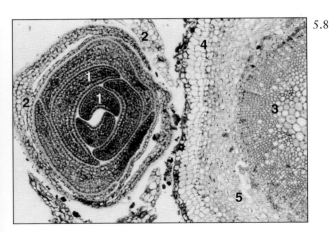

5.8

5.8 TS of the shoot of the dicotyledon *Ligustrum vulgare* (privet) showing a dormant axillary bud. The decussately-arranged foliage leaf primordia (1) are invested by scale leaves (2). The stem has some secondary xylem (3) and cork (4) has already developed. Cortex (5). (LM x 70.)

1	Foliage leaf primordia	4	Cork
2	Scale leaves	5	Cortex
3	Secondary xylem		

5.9

5.9 Stump of an old shoot of the monocotyledon *Musa* (banana). The stem (1) is ensheathed by the bases of the large foliage leaves; the numerous parallel and longi-tudinally-orientated fibre strands in the sheaths greatly strengthen the inflorescence axis which at maturity reaches several metres in length and bears at its tip a very heavy crop of bananas.

1	Stem

5.10

5.10 Trunk of the monocotyledon *Archontophoenix alexandrae* (Alexander palm). Note the compound nature of the large leaf blade whose petiole (1) expands into a leaf sheath (2) encircling the trunk.

1	Petiole	2	Leaf sheath

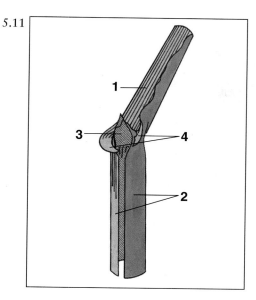

5.11

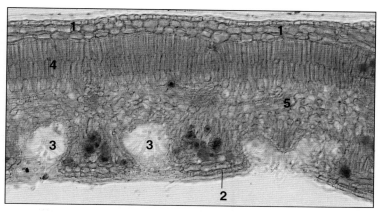

5.12

5.11 Diagram of a typical grass leaf. At the junction of the leaf blade (1) with the sheath (2) a median membranous ligule (3) and a pair of lateral auricles (4) are present.

1	Leaf blade
2	Sheath
3	Membranous ligule
4	Auricles

5.12 TS of the lamina of the bifacial leaf of the dicotyledon *Nerium oleander*. In this xeromorphic species stomata are absent from the adaxial surface and the multiple epidermis (1) is covered by a very thick cuticle. The single abaxial epidermal layer (2) has a thinner cuticle and its stomata are confined to hair-lined crypts (3). The mesophyll is differentiated into adaxial palisade (4) covering a layer of spongy tissue (5). (LM x 110.)

1	Adaxial multiple epidermis	4	Palisade mesophyll
2	Abaxial epidermis	5	Spongy tissue
3	Hair-lined crypts		

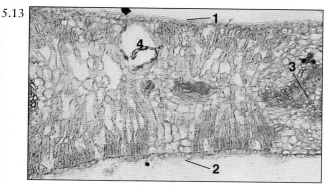

5.13

5.13 TS of the isobilateral leaf of the dicotyledon *Eucalyptus*. Palisade mesophyll occurs at both the adaxial (1) and abaxial (2) surfaces. The xylem (3) is adaxial. Oil gland (4). (LM x 90.)

| 1 | Adaxial surface | 3 | Xylem |
| 2 | Abaxial surface | 4 | Oil gland |

5.14

5.14 TS of the leaf of the monocotyledon *Iris* showing its unifacial blade (**A**) and bifacial sheath (**B**). Note that in the numerous veins the phloem (blue) lies nearest the abaxial surface, whilst the xylem (red) faces the adaxial surface (**B**) or lies towards the confluent adaxial surfaces (**A**).

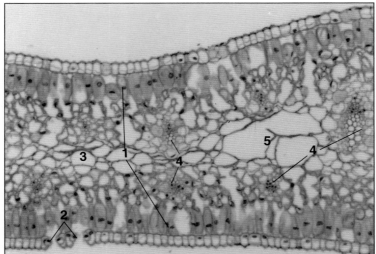

5.15 TS of the unifacial leaf blade of the monocotyledon *Narcissus* (daffodil). Note the single layer of palisade mesophyll (1) underlying the abaxial epidermis and the numerous stomata (2). Within the central layer of non-photosynthetic parenchyma (3) two series of veins occur, with the xylem (4) of opposed series facing towards each other (cf., **5.14**). Mucilage cavity (5). (LM x 100.)

1	Palisade mesophyll
2	Stomata
3	Non-photosynthetic parenchyma
4	Xylem
5	Mucilage cavity

5.16

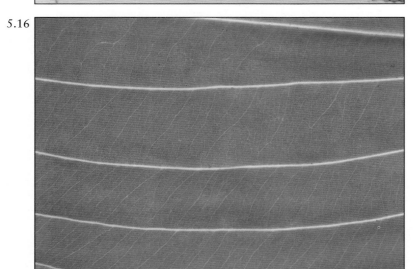

5.16 Leaf of the monocotyledon *Hosta* showing its venation. The main parallel longitudinal veins are connected by abundant, obliquely transverse, branched commisures.

5.17

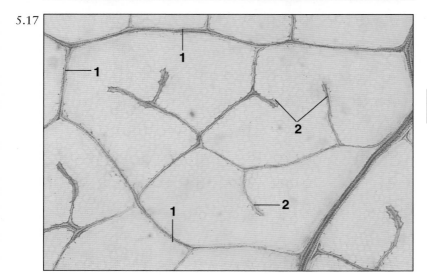

5.17 Cleared dicotyledonous leaf showing the xylem and its reticulate venation. Note the demarcation of areoles in the mesophyll which are enclosed by veinlets (1) whose branches end blindly (2). (LM x 120.)

1	Veinlets bounding an areole
2	Blindly-ending veinlets

5.18 Section parallel to the surface of the leaf of the dicotyledon *Ligustrum vulgare* (privet). Note the prominent protoxylem elements in the veinlets (cf., **5.17**) and the numerous air spaces in the mesophyll of the areoles. (Phase LM x 285.)

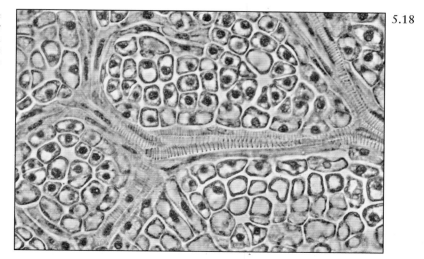

5.18

5.19 Leaf of the dicotyledon *Magnolia* showing its venation. The pinnately-arranged laterals are interconnected by smaller tertiary veins which further branch to form a reticulum (cf., **5.17**).

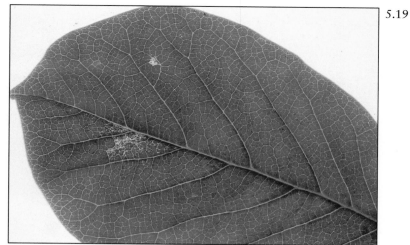

5.19

5.20 Fossilised remains of a compound leaf of the cycad *Nilsonia*. Note the parallel venation of the leaflets; these veins show some dichotomous branching.

5.20

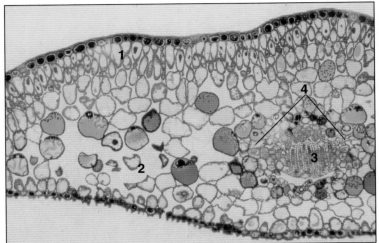

5.21

5.21 TS of the bifacial evergreen conifer leaf of *Taxus baccata* (yew). The mesophyll consists of the tightly-packed, elongate cells of the adaxial palisade (1) whilst the spongy tissue (2) shows extensive intercellular air spaces. The single median vein (3) is un-branched but transfusion tracheids (4) extend laterally into the mesophyll. Note the numerous stomata (cf., **1.15**) and the cuticular papillae in the abaxial epidermis. (G-Os, LM x 70.)

1	Adaxial palisade
2	Spongy tissue
3	Median vein
4	Transfusion tracheids

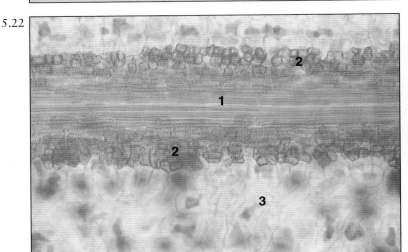

5.22

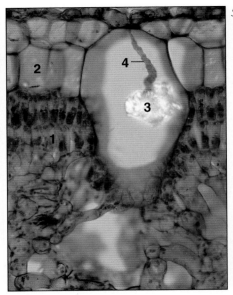

5.2

5.22 Cleared leaf of the conifer *Taxus baccata* (yew) showing detail of its midrib. Note the median vein (1) and the transfusion tracheids (2) on its lateral margins (cf., **5.21**) which extend into the spongy mesophyll (3). (LM x 110.)

1	Median vein
2	Transfusion tracheids
3	Spongy mesophyll

5.23 TS showing detail of the bifacial leaf of the dicotyledon *Ficus elastica* (rubber-fig plant). The palisade mesophyll (1) is adaxially covered by a layer of large water storage cells (2) which represent the inner layers of a multiple epidermis derived by periclinal divisions of the protoderm. Note the cystolith which consists of a precipitate of calcium carbonate (3) terminating a stalk of cellulose (4) connected to the wall of an enlarged epidermal cell. (Polarised LM x 260.)

1	Palisade mesophyll
2	Water storage cells
3	Cystolith
4	Cellulose stalk

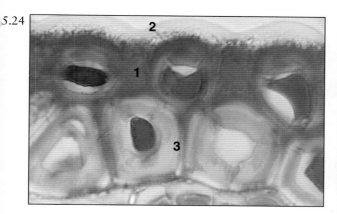

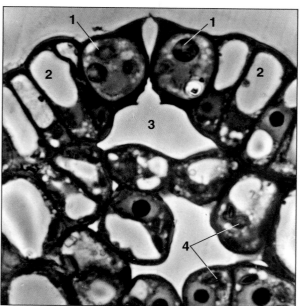

5.24 TS showing surface detail of the xeromorphic leaf of the conifer *Pinus monophylla* (pine). Note the thick-walled, lignified epidermal cells (1) which are coated externally by a thick cuticle (2) and the hypodermal sclerenchyma (3). (LM x 900.)

1	Epidermal cells
2	Cuticle
3	Hypodermal sclerenchyma

5.25 TS of the young stem of the dicotyledon *Phaseolus vulgaris* (bean) showing detail of a stoma. The guard cells have dense cytoplasm and prominent amylochloroplasts (1) whereas the ordinary epidermal cells have large vacuoles (2) and are without chloroplasts. The anticlinal guard cell walls adjacent to the stomatal pore and the periclinal walls are thickened, and the outer periclinal walls extend into prominent ledges over the pore. Note the sub-stomatal space (3) and the chloroplasts (4) in the mesophyll. (G-Os, x 1,100.)

1	Amylochloroplasts	3	Sub-stomatal space
2	Vacuoles	4	Mesophyll chloroplasts

5.26 TS of the abaxial surface of the leaf of the monocotyledon *Clivia miniata.* The epidermis is covered by a thick cuticle (1) which is also present over the guard cell (2). The spongy mesophyll shows large intercellular spaces (3). (G-Os, x 300.)

1	Cuticle
2	Guard cell
3	Intercellular spaces

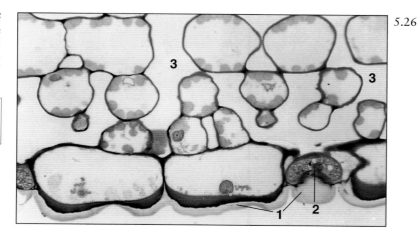

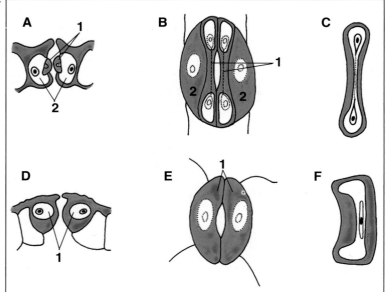

5.27 Diagrammatic representations of stomata from a grass (**A–C**) and a dicotyledon (**D–F**). **B** and **E** show surface views, **A** and **D** are transverse sections while **C** and **F** are cut longitudinally through a guard cell. In the grass the dumbbell-shaped guard cells (1) have unevenly thickened walls, and are dwarfed by the larger subsidiary cells (2). The dicotyledon illustrated lacks subsidiary cells (cf., 5.30); note the kidney-shaped guard cells (1) with unevenly thickened walls.

1 Guard cells
2 Subsidiary cells

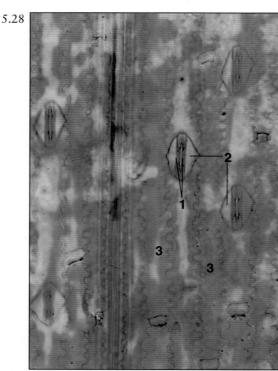

5.28 Stomata in a surface view of a cleared leaf blade of the monocotyledon of *Zea mays* (maize). Note that the long axes of the guard cells (1), the subsidiary cells (2) and the ordinary epidermal cells (3, with sinuous walls) all lie parallel to the axis of the leaf. (LM x 305.)

1 Guard cells
2 Subsidiary cells
3 Epidermal cells

5.29 Abaxial leaf surface of the dicotyledon *Begonia rex*. Note the numerous, large multicellular trichomes (1) and stomata (2). (SEM x 80.)

1 Multicellular trichomes
2 Stomata

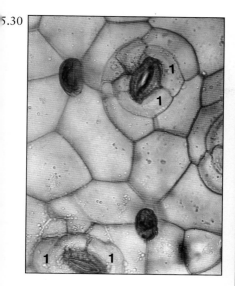

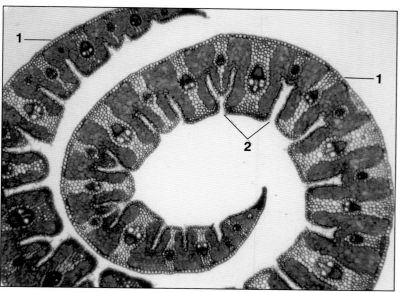

5.30 Abaxial surface view of the leaf of the dicotyledon *Peperomia argyreia*. Note the subsidiary cells (1) and that the long axes of the stomata lie in varying orientations to each other (cf., **5.28**). (LM x 225.)

1	Subsidiary cells

5.31 TS of the xeromorphic, bifacial leaf blade of the monocotyledon *Ammophila arenaria* (marram grass). In life the leaf blade is variably rolled, according to the humidity of the atmosphere and availability of water, with the smooth abaxial surface (1) outermost, while the inrolled adaxial surface has longitudinal grooves (2, cf. **5.32**). (LM x 30.)

1	Abaxial surface
2	Adaxial grooves

5.32 TS of *Ammophila arenaria* (marram grass) showing detailed structure of the lamina. The abaxial epidermis (1) of this monocotyledon is covered by a thick cuticle and lacks stomata. A hypodermal layer of sclerenchyma (2) extends into the ridges (3) on the adaxial surface. The mesophyll (4) is confined to the adaxial surface, while numerous long hairs extend from the epidermis. The adaxial cuticle is thin. The stomata occur on the margins of the adaxial ridges and in the grooves conspicuous hinge cells are present (5). (LM x 120.)

1	Abaxial epidermis	4	Mesophyll
2	Sclerenchyma	5	Hinge cells
3	Adaxial ridges		

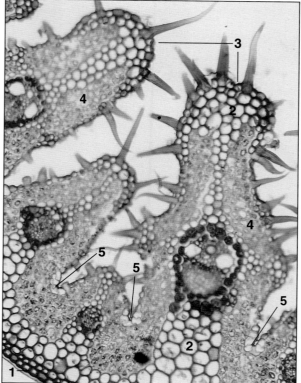

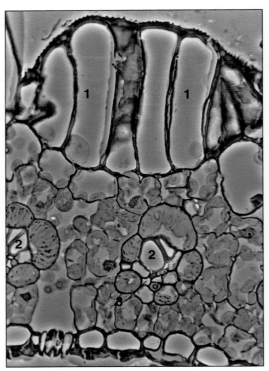

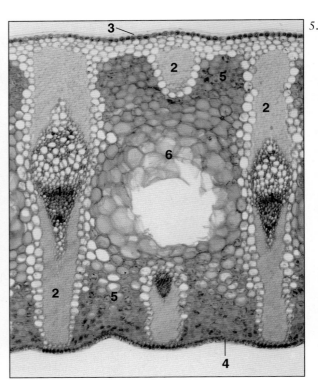

5.33 TS of bulliform cells in the lamina of the grass *Zea mays* (maize). The large, thin-walled and highly vacuolate bulliform cells (1) of this monocotyledon are confined to the adaxial epidermis. When they lose water, they contribute to the rolling of the leaf. Tracheary elements (2), sieve tubes (3). (Phase LM x 495.)

1	Bulliform cells
2	Tracheary elements
3	Sieve tube

5.34 TS of the isobilateral leaf of the monocotyledon *Phormium tenax* (New Zealand flax). The veins (1) are embedded in thick-walled, heavily lignified fibres (2) which form a series of 'girders' linking ad- (3) and abaxial (4) surfaces. The fibres are used commercially for cordage and individual fibres may reach up to 15 mm in length. Mesophyll (5), non-photosynthetic parenchyma (6). (LM x 80.)

1	Veins	4	Abaxial surface
2	Lignified fibres	5	Mesophyll
3	Adaxial surface	6	Non-photosynthetic parenchyma

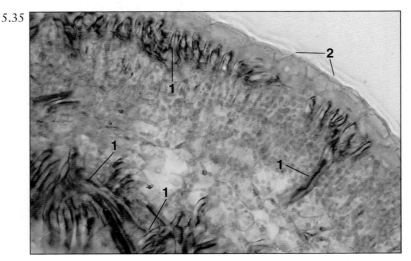

5.35 TS of the xeromorphic leaf of the dicotyledon *Olea sativa* (olive). Note the numerous thick-walled sclereids (1) ramifying in the mesophyll and the very thick cuticle (2) coating the adaxial epidermis. (LM x 120.)

1	Sclereids
2	Cuticle

5.36 TS of the midrib of the bifacial leaf
of the dicotyledon *Prunus laurocerasus*
(cherry laurel). The lamina consists of a
compact adaxial palisade mesophyll (1) and
an irregular spongy layer adaxially (2).
Both surfaces are covered by a conspicuous
cuticle. The midrib shows a single large vein
with adaxial xylem (3), whilst a strand of
collenchyma (4) causes a slight ridge on the
adaxial surface. Cambial-like layer of
parenchyma (5), phloem (6). (LM x 80.)

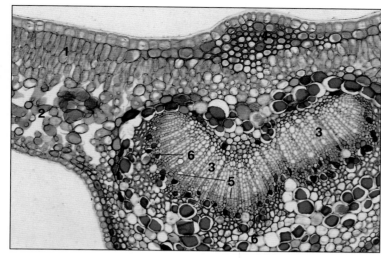

1	Adaxial palisade mesophyll
2	Spongy layer
3	Xylem
4	Collenchyma strand
5	Cambial-like layer
6	Phloem

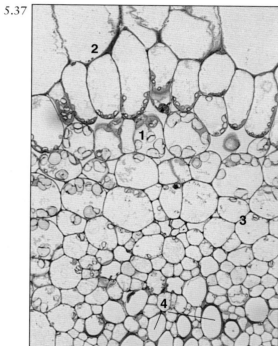

5.37 TS showing detail of the succulent lamina of the
dicotyledon *Peperomia resedaeflora*. This rain forest
epiphyte shows only a single layer of palisade mesophyll
(1) which is covered adaxially by large, highly vacuolated
water storage cells (2). These form a layer over half the
total thickness of the leaf, but only the innermost storage
cells are shown in this illustration. Underlying the palisade
cells compact parenchyma cells (3) contain large
amylochloroplasts. Xylem (4). (G-Os, LM x 285.)

1	Palisade mesophyll	3	Parenchyma cells
2	Water storage cells	4	Xylem

5.38 Periclinal chimaeral leaves of the dicotyledon
Pelargonium (geranium). Their variegated appearance is
due to a mutation in a single corpus cell at the extreme tip
of the shoot apex; derivatives of such a cell contain
proplastids which are incapable of differentiating into
chloroplasts. The mesophyll of the leaf primordium
develops from the inner tunica layer(s) and corpus but only
the former are capable of giving rise to green cells:
proliferation of this tissue at the leaf margins gives a green
border to the leaf.

111

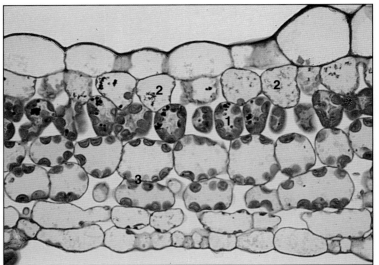

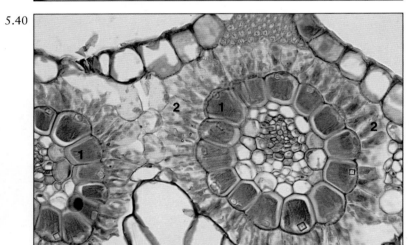

5.39 TS showing detail of a variegated bifacial leaf of the dicotyledon *Glechoma hederacea* (ground ivy). In this chimaeral leaf the photosynthetic palisade mesophyll is reduced to a single layer (1) of squat cells whilst the hypodermal palisade layer (2) is devoid of chloroplasts. The abaxial spongy mesophyll (3) contains chloroplasts but is more compact than in the non-variegated leaf (cf. **3.3**). Note the absence of chloroplasts from both epidermises. (G-Os, LM x 310.)

1	Photosynthetic palisade mesophyll
2	Hypodermal non-photosynthetic palisade
3	Spongy mesophyll

5.40 TS of the leaf of the grass *Panicum turgidum*. This desert species shows typical 'Kranz' anatomy, with the parenchyma cells of the prominent bundle sheath (1) containing large aggregated chloroplasts. The mesophyll cells (2) radiate out from the sheath and contain smaller, discrete chloroplasts. (LM x 310.)

1	Bundle sheath
2	Mesophyll cells

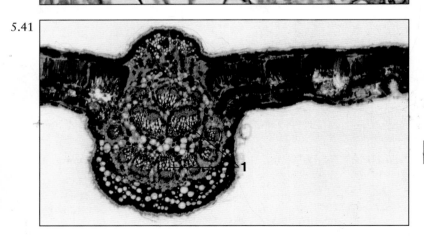

5.41 TS of the bifacial leaf of the dicotyledon *Banksia*. In this indigenous Australasian genus the xeromorphic leaves are covered on both surfaces by a thick cuticle and much tannin is present in the mesophyll. In the midrib a number of discrete veins are present and the orientation of their xylem strands (1) is variable. (LM x 10.)

1	Xylem strands

5.42 TS of a large vein from the leaf of the monocotyledon *Dracaena draco* (dragon's blood). Vascular cambium is lacking between the adaxially situated xylem (1) and abaxial phloem (2). The vein is partially separated from the mesophyll (3) by a sheath of fibres (4) (G-Os, LM x 295.)

1 Adaxial xylem	3 Mesophyll
2 Abaxial phloem	4 Fibres

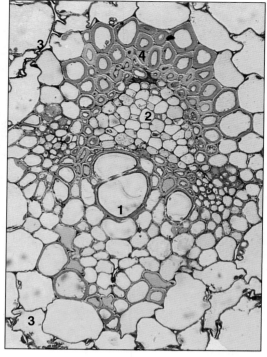

5.43 TS of the main vein from the leaf of the dicotyledon *Sorbus aucuparia* (rowan). The adaxially situated xylem shows wide-diametered empty tracheary elements whose secondary walls bear bordered pits (1). Several, narrower, differentiating elements (2) are also evident with thickened secondary walls but intact protoplasts. A cambial zone (3) separates the xylem from the phloem fibres (4). (G-Os, TEM x 2,100.)

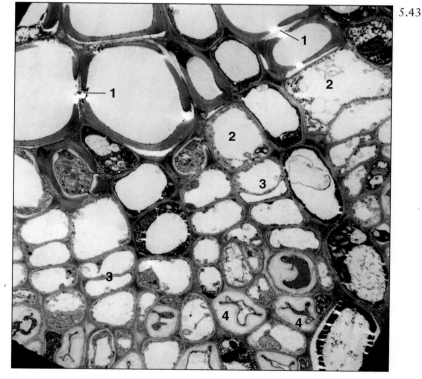

1 Bordered pits	3 Cambial zone
2 Differentiating xylem	4 Phloem fibres

5.44

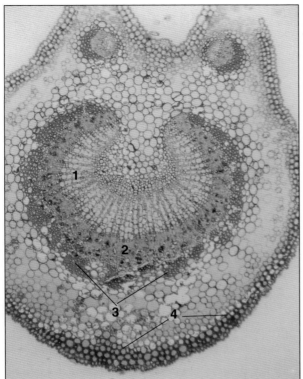

5.44 TS of the petiole of the dicotyledon *Phaseolus vulgaris* (bean). A large crescent-shaped vein and a pair of small lateral veins are present. The extensive xylem (1) of the large vein lies adaxially whilst the phloem (2) is delimited by strands of fibres (3). An extensive hypodermal band of sclerenchyma (4) is evident and the remaining ground tissue is parenchymatous. (G-Os, LM x 85.)

1	Xylem	3	Fibre strands
2	Phloem	4	Sclerenchyma

5.45

5.46

5.45, 5.46 Potted specimen of the dicotyledon *Oxalis angularis* showing sleep movements. In the day-time (**5.45**) the three leaflets of each leaf are extended but they droop at night (**5.46**). This movement is caused by loss of turgor in specialised tracts of parenchyma tissue located in a joint-like thickening (pulvinus) situated at the top of the petiole just beneath the leaflets.

5.47 LS of the cotyledonary node of the dicotyledon *Phaseolus vulgaris* (bean). Germination is epigeal and the cotyledons wither and fall off within several weeks. Note the axillary bud (1) between the stem and the cotyledon (2); a well-developed abscission zone (3) runs across the cotyledon base and traverses the main cotyledonary vein (4). (G-Os, LM x 75.)

1	Axillary bud
2	Cotyledon
3	Abscission zone
4	Cotyledonary vein

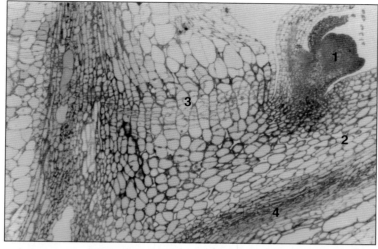

5.47

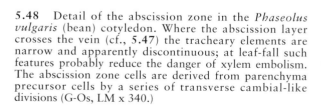

5.48

5.49

5.48 Detail of the abscission zone in the *Phaseolus vulgaris* (bean) cotyledon. Where the abscission layer crosses the vein (cf., **5.47**) the tracheary elements are narrow and apparently discontinuous; at leaf-fall such features probably reduce the danger of xylem embolism. The abscission zone cells are derived from parenchyma precursor cells by a series of transverse cambial-like divisions (G-Os, LM x 340.)

5.49 Detail of a corky leaf scar on the stem of the monocotyledon *Monstera*. Note also the smaller vein scar (1).

1	Vein scar

5.50 Base of the mature trunk of the palm *Corphyra elata*. Note the spiny leaf bases which persist for many years after the withered fronds of this monocotyledon have been removed.

CHAPTER 6

The stem

Primary growth

The aerial stem bears the green photosynthetic leaves and the reproductive organs (**1.16**, **4.13**, **6.1**, **6.2**), while underground stems are frequently perennating and food storage organs (**6.3**). Most unthickened stems are cylindrical (**6.2**), but ridged and rectangular forms (**6.4**, **6.5**) are common. Stems are sometimes flattened, leaf-like structures (phylloclades, **6.6**, **6.7**) with their leaves reduced to scales. The slender xeromorphic stem of *Casuarina* (**6.8**) bears only scale leaves, so that photosynthesis occurs in the stem cortex which lies adjacent to the longitudinal, hair-lined grooves.

In succulents the stems are swollen and also photosynthetic and in many species of cacti and spurges (*Euphorbia*) the leaves are represented by spines (**3.33**, **6.9**). Starch is commonly stored in the parenchymatous ground tissue of the stem (**2.31**) and is particularly abundant in the swollen stems of succulents and the underground stems of corms, tubers and rhizomes (**6.3**). On the condensed shoots of rosette species (**4.13**) the leaves are crowded and the internodes short, but at flowering the internodes commonly become much more widely spaced as is dramatically shown in *Agave* (**6.2**).

Anatomy of the mature primary stem

The vascular system in the young internode usually consists of separate vascular bundles (**6.4**, **6.5**, **6.10**, **6.11**) that typically form a peripheral cylinder in dicotyledons (**1.28**) but are scattered in monocotyledons (**1.27**). The cortex lies external to the vascular tissue and is bounded by an epidermis which often bears stomata and trichomes (**5.25**, **6.5**, **6.8**). The ground tissue in which the vascular tissue is embedded is basically parenchymatous and the cortex is often photosynthetic (**6.6**, **6.8**, **6.11**). In dicotyledons a parenchymatous pith is usually present, but vascular bundles occasionally may be present centrally (**6.10**). In the majority of monocotyledons the bundles occur throughout the ground tissue (**1.27**), but sometimes a pith is

present (**6.4**). Sclerenchyma fibres are often present in the ground tissue (**6.4**, **6.8**, **6.10**) and the parenchyma may become lignified. Collenchyma frequently occurs just beneath the epidermis, especially at the angles of the stem (**6.5**). In some stems a prominent starch sheath occurs in the innermost cortical layer and in underground stems this may develop thickening to form an endodermis (**6.12**).

Vascular bundles in the stem are commonly collateral, with the phloem lying nearest to the epidermis and the xylem situated internally and on the same axis (**1.11**, **6.6**, **6.10**). Bicollateral bundles may also occur in which the phloem lies both external and internal to the xylem (**6.11**). In many monocotyledons the bundles are amphivasal with a central strand of phloem surrounded by xylem (**6.12**). Amphicribral bundles, in which the xylem is surrounded by phloem (**6.13**), occur in ferns and a few angiosperms, while in others the bundles may lack xylem. In the great majority of dicotyledonous stems, a cambial layer is located between the xylem and phloem (**6.6**, **6.11**) but in monocotyledons this is absent (**1.11**, **6.12**).

The vascular anatomy at the node is more complex than in the internode due to the branches that pass outwards from the axial vascular bundles to the leaves and axillary branches (**4.17**, **4.18**, **4.23**, **6.14**). Apart from branching at the nodes, the axial bundles normally interconnect with adjacent vascular bundles at various levels along the internodes. In monocotyledons, axial bundles often run obliquely for some distance in the internode and have frequent interconnections (**6.15**) with numerous veins (leaf traces) passing outwards to each leaf. In dicotyledons there are usually fewer leaf traces. In species with few interconnections between axial bundles, damage to one part of the axial system may severely disrupt the supply of water and nutrients to parts of the plant lying above or below the injury site.

In most stems the protoxylem and protophloem elements (**2.3**) are damaged during elongation and expansion growth (**1.11**), so that in the older primary stem (**3.5**, **6.10**, **6.11**) only the metaxylem and metaphloem are normally functional. Pericyclic fibres often develop in the outer procambium (**1.4**, **3.2**) and replace the crushed, isolated files of protophloem, while the protoxylem is sometimes represented by lacunae (**1.11**) after the 'holey' primary walls (**2.66**) of the tracheary elements become over-extended. In dicotyledons the metaxylem vessels are frequently arranged in radial files separated by parenchyma or sclerenchyma (**3.5**). In monocotyledons the relatively few vessels are usually larger and parenchyma or sclerenchyma often occurs between them (**1.11**).

Modifications of the primary stem

In submerged aquatic stems the epidermis usually lacks both a cuticle and stomata. Chloroplasts often occur in the epidermis and are normally abundant throughout the aerenchymatous cortex (**6.16**). The vascular tissue is concentrated in a central cylinder with the xylem greatly reduced and restricted to annular or spiral tracheids whilst protoxylem lacunae are common (**6.17**). Desert perennials are commonly succulent and store water in their pith and cortex. They have small or vestigial leaves (**3.33**, **6.9**) and the stem cortex is the main site of photosynthesis. The epidermis is often multiseriate and is covered by a thick cuticle but a transparent cork layer may develop in *Euphorbia* and some cacti. In plants growing in salt marsh or other saline environments the leaves are also often reduced and the stems are succulent.

Secondary growth

Most dicotyledons and all gymnosperms undergo some degree of secondary thickening (**1.30**, **3.2**, **3.27**, **3.44**). The amount of thickening produced depends upon whether the mature plant is herbaceous (**6.18**, **6.19**) or arborescent (**1.2**, **3.44**). The fascicular vascular cambium develops from a narrow strip of procambium between the xylem and phloem which remain meristematic after the primary vascular tissues have matured in a bundle. At the onset of cambial activity the divisions are normally localised within the individual vascular bundles (**1.28**, **3.5**, **6.18**). They then spread laterally through the adjacent interfascicular parenchyma cells so that a continuous cylinder of vascular cambium eventually results (**1.30**). The vascular cambium normally commences activity by the end of the first season's growth.

Anatomy of the woody stem

The vascular cambium consists of fusiform and ray initials (**6.20**, **6.21**). The cambium is often storeyed, with the fusiform initials arranged in approximately horizontal layers when viewed in tangential longitudinal section (**6.21**). However, in a non-storeyed cambium (**6.21**) the fusiform initials tend to be longer and their end walls taper more acutely than in the storeyed cambium. The fusiform initials give rise to the axial components of the woody stem: vessels, tracheids, fibres, sieve tubes, companion cells and axial parenchyma cells. These may be storeyed or otherwise (**3.63**) according to the pattern of the cambium from which they are formed. The tangential (periclinal) walls of the fusiform initials are wider than the radial walls (**6.20**). The initials divide tangentially (**6.20**), cutting off xylem centripetally and phloem centrifugally (**3.27**, **6.22**). During the growing season the cambial initials are actively dividing. They are highly vacuolate cells and the expanding cell plate formed after mitosis is invested at its periphery by a prominent phragmoplast (**2.61**, **6.20**, **6.22**).

Each cambial initial produces radial rows of derivatives and in an active cambium a fairly wide cambial zone is apparent (**6.18**). In this zone tangential divisions also occur within the potential xylem and phloem elements (**6.22**). In gymnosperm and some angiosperm xylem the radial seriation is retained as the tracheary elements mature (**3.17**, **3.53**, **6.23**). In most angiosperms this radial pattern is more or less severely disturbed by the maturation of large-lumened vessels (**3.67**, **3.68**, **6.18**). The ray initials are approximately isodiametric and they divide tangentially (**6.20**) to form the rays which run radially across the secondary vascular system (**3.17**, **3.44**). To accommodate the increasing circumference of the stem, as secondary thickening progresses, the fusiform initials sometimes divide radially (anticlinally) to form additional fusiform initials. They also give rise to further initials which form new rays to meet the increased requirements for radial transport of water and nutrients in the expanding stem.

The expansion of the stem brought about by secondary growth is accompanied by various changes. The primary phloem and xylem cease to function in translocation and transpiration. The pith often remains more or less intact over a number of years but may breakdown to form a central cavity. The cortex may persist for some time (3.17). The primary phloem and the early secondary phloem (3.27, 6.23) frequently become crushed between the pericyclic or cortical sclerenchyma and the expanding xylem cylinder, and even in older stems the secondary phloem remains a relatively narrow layer (3.44). However, if axial fibres develop abundantly in the secondary phloem the older tissue may remain discrete (3.37). To accommodate the increasing circumference of the stem, the intervening ray parenchyma cells divide periclinally so that the rays flare outwards (3.37).

In a number of secondary thickened dicotyledonous stems, especially in climbing plants (lianes), the pattern of vascular development is anomalous. In the tropical liane *Bauhinia* the older stem (6.24) is irregular and often ribbon-like and somewhat flexible. In early secondary growth the cambium produces uniform xylem. However, the cambium soon becomes more active on opposing sectors of the stem and produces thick wedges of large-lumened vessels (6.25), while the intervening areas of xylem are thin and have narrow tracheary elements. The form of the mature liane becomes more complicated as further vascular tissues differentiate from the dilated rays running across the fissured xylem (6.24). In *Tinospora* the rays are greatly inflated so that the secondary xylem is separated into tapering wedges (6.26). In other stems new vascular cambia may arise successively further from the centre of the stem producing concentric rings of xylem and phloem, or the cambium sometimes produces phloem centripetally.

Periderm

In gymnosperms and woody dicotyledons the epidermis of the stem is normally replaced by the protective cork and associated tissues (3.13, 3.17, 3.41, 3.44, 6.27, 6.28). The outermost region of this periderm comprises the phellem (cork, 1.30) derived from the phellogen (cork cambium). This is a meristematic layer of tangentially flattened cells which commonly arises hypodermally (5.8). It may also be of epidermal origin (6.29) or may form deeper in the cortex. In some species the cork cambium cuts off a little parenchymatous tissue (phelloderm) internally. The cork consists of radially aligned, tightly packed cells; they are dead and have thickened walls which are suberised and impermeable. Periderm also forms over wounded surfaces (5.48, 5.49, 6.30) and occurs in thickened stems of some monocotyledons (6.31).

Commercial cork is harvested from *Quercus suber* (3.13) and is stripped off the trees in cycles of about 10 years (6.28). In this species each phellogen produces several millimetres of cork and is then replaced by new phellogens which arise in successively deeper regions of the cortex and eventually in the outer secondary phloem (3.41). The removal of the cork crop does not harm the vascular cambium of the oak tree. In most woody species the outer dead bark (rhytidome) is periodically sloughed off the trunk and main branches. The rhytidome consists of successively deeper-formed, discontinuous but overlapping periderms and intervening patches of non-functional phloem (3.41).

The cork is impermeable to gaseous diffusion but the numerous lenticels (4.22, 6.32, 6.33) facilitate the movement of oxygen into the living tissues within this barrier and also allow the exit of carbon dioxide. Lenticels arise from less tightly-packed regions of the phellogen, and the cork (complementary tissue) produced consists of rounded cells with large intracellular spaces between them. In many woody species layers of more compact cork are produced periodically and these retain the loose complementary cells within the lenticel.

Thickened monocotyledonous stem

The majority of monocotyledons are herbaceous, but in bamboos and some other species the stem is relatively wide (6.34) due to the activity of the primary thickening meristem. In this meristem the cells are aligned in a transverse or oblique sheet and undergo periclinal divisions (6.15), with the internal derivatives differentiating into the axial vascular bundles and ground parenchyma (4.15). During early growth of most palms the internodes remain short, while diffuse growth and division within the ground parenchyma leads to the stem becoming progressively thicker. When the stem reaches its adult diameter (5.50), internodal elongation occurs and some species may attain great heights and life span.

In the very few truly woody monocotyledonous genera (e.g. *Dracaena*, *Yucca*, *Cordyline*) anomalous secondary growth occurs because of the activity of the secondary thickening meristem (1.22, 5.2, 6.35). This forms in the outer cortex of the older stem (6.36) and often links with the primary thickening meristem of the younger shoot. The secondary thickening meristem cuts off complete vascular bundles centripetally which are frequently amphivasal (**6.36**) in contrast to the collateral bundles in the primary stem. In other arborescent monocotyledons (palms, pandans, **1.26**, **5.10**) a secondary thickening meristem is absent but diffuse secondary growth occurs due to divisions throughout the ground parenchyma of the stem.

6.1 Detail of the aerial shoot of the monocotyledon *Musa* (banana). Note the numerous parallel veins in the leaves (cf., **5.9**). Massive whorls of immature seedless fruits (developed without fertilisation) are borne on a hanging inflorescence stalk. Near its tip lies a whorl of flowers (1); their basal ovaries are beginning to enlarge and later the withered flowers will be discarded. At the tip of the inflorescence a large bract (2) hides clusters of sterile male flowers.

6.2 Flowering shoots of the monocotyledon *Agave americana* (century plant). This perennial monocotyledon grows vegetatively as a rosette bearing numerous sword-shaped, xeromorphic leaves. However, after many years vegetative growth reproduction occurs and an inflorescence axis (several metres tall) arises with numerous small bracts, in whose axils a number of flowers are borne on short lateral branches.

| 1 | Whorl of flowers |
| 2 | Bract |

6.3 Swollen underground stem tuber of the dicotyledon *Solanum tuberosum* (potato). This perennating organ is almost entirely composed of compact, large parenchyma cells containing considerable quantities of starch (cf., **2.31**, **2.32**). In nature the thin stolons that connect the tubers to the mother plant die off and the isolated tubers give rise to daughter plants. Note the numerous sprouting axillary buds (several of which arise from each 'eye' of the tuber) bearing leaf primordia (1) at their tips. Adventitious root primordia (2).

| 1 | Leaf primordia |
| 2 | Root primordia |

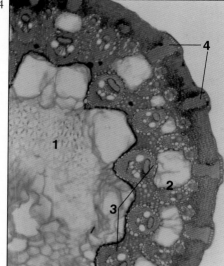

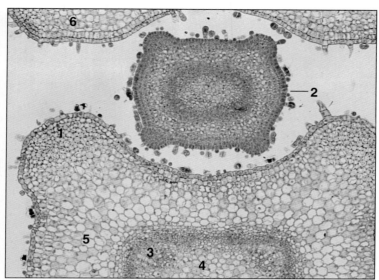

6.4 TS of the hydrophytic stem of *Juncus communis* (rush). This shows a wide pith (1) and large air cavities occur in the outer pith and cortex (2). Unlike the majority of monocotyledons the vascular bundles (3) are peripherally distributed; however the absence of a vascular cambium between the xylem and phloem distinguishes this stem from that of a dicotyledon. Fibre bundles (4). (LM x 65.)

1	Pith
2	Cortex
3	Vascular bundles
4	Fibre bundles

6.5 TS of a young node of the dicotyledon *Glechoma hederacea* (ground ivy). The corners of its rectangular stem are swollen by peripheral collenchyma (1) and the axillary bud (2) shows a similar form. Vascular tissue (3), pith (4), cortex (5), subtending leaf (6). (LM x 80.)

1	Collenchyma	4	Pith
2	Axillary bud	5	Cortex
3	Vascular tissue	6	Subtending leaf

6.6 TS of the phylloclade of *Muehlenbeckia platyclados*. In this plant the flattened stem acts as a photosynthetic organ but the peripheral distribution of vascular bundles containing vascular cambium (1), demonstrates that this is a dicotyledonous stem. Photosynthetic cortex (2), pith (3). (LM x 80.)

1	Vascular cambium
2	Photosynthetic cortex
3	Pith

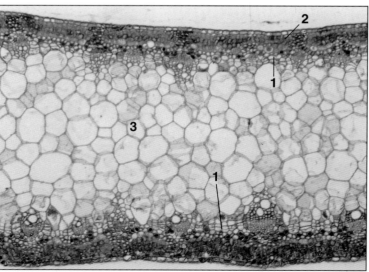

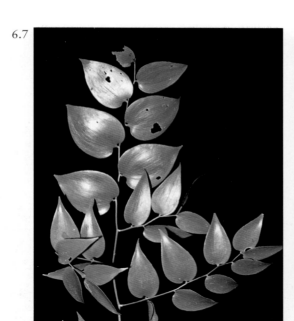

6.7

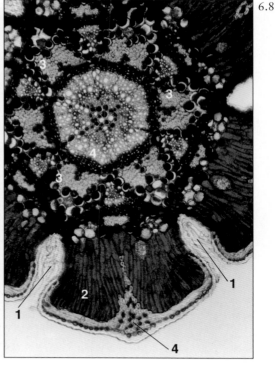

6.8

6.7 Shoot of the monocotyledon *Semele* showing its green leaf-like phylloclades. These represent flattened shoots of limited growth which are developed from buds borne in the axils of scale leaves situated on the stems (1) which are cylindrical.

1	Main stem

6.8 TS of the stem of the dicotyledon *Casuarina*. The stem of this nitrogen-fixing plant bears only scale leaves and the photosynthetic function of the plant is assumed by the xeromorphic stem. This shows hair-lined grooves (1) in which stomata occur, and chlorenchyma (2) at the margins of the grooves. The epidermis has a thick cuticle and tracts of sclerenchyma fibres (3) occur both hypodermally and internally. Note the ring of vascular bundles (4). (LM x 115.)

1	Hair-lined grooves	3	Sclerenchyma fibres
2	Chlorenchyma	4	Vascular bundles

6.9

6.9 Several large specimens of *Echino-cactus*. In this xeromorphic dicotyledon the highly modified leaves occur as spines whilst the green succulent stem is the photosynthetic organ. Each vertical ridge on the stem shows a row of tubercles which represent axillary buds bearing several spines.

6.10 TS of the young stem of the dicotyledon *Piper*. This shows an outer ring of variably-sized vascular bundles, demarcated internally by a sheath of sclerenchyma (1), whilst in the pith a number of medullary vascular bundles occur (2). Such diffuse distribution of bundles is uncommon in dicotyledons (cf., **1.28**). (LM x 85.)

1	Sclerenchyma sheath
2	Medullary vascular bundles

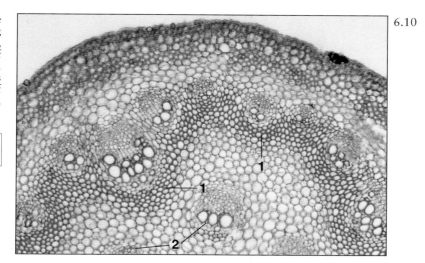
6.10

6.11 TS of a vascular bundle in the stem of *Trichosanthes*. This dicotyledon shows bicollateral bundles with strands of external (1) and internal phloem (2) on either side of the xylem. Note the very wide xylem vessels (3) and the conspicuous sieve tube with sieve plates (4). Vascular cambium (5), sclerenchyma (6), photosynthetic cortex (7). (LM x 75.)

1	External phloem
2	Internal phloem
3	Xylem vessels
4	Sieve plates
5	Vascular cambium
6	Sclerenchyma
7	Photosynthetic cortex

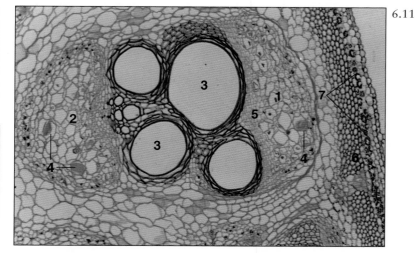
6.11

6.12 TS of the rhizome of the mono-cotyledon *Convallaria majalis*. A one- to several-layered and uniformly thickened, endodermis (1) delimits the wide cortex (2) from the pith in which occur scattered vascular bundles. The inner bundles are amphivasal with xylem (3) surrounding the phloem (4). (LM x 115.)

1	Endodermis
2	Cortex
3	Xylem
4	Phloem

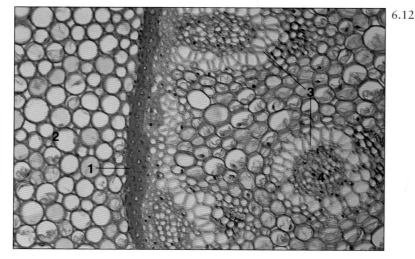
6.12

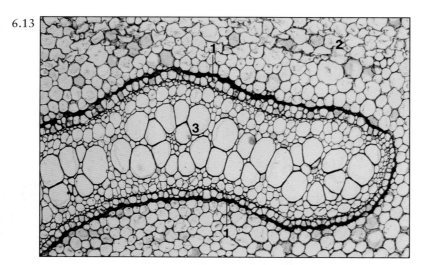

6.13   TS of the rhizome of the fern *Pteridium aquilinum* (bracken). An extensive vascular strand is separated by an endodermis (1) from the ground parenchyma (2). The large-diametered xylem elements (3) are surrounded by smaller phloem parenchyma and sieve cells. (LM x 85.)

1	Endodermis
2	Ground parenchyma
3	Xylem elements

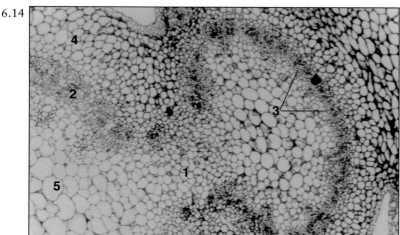

6.14   TS of the young node of the dicotyledon *Phaseolus vulgaris* (bean). Note the gap (1) in the vascular cylinder of the stem (2) where it branches to supply the vascular system (3) of the axillary bud. Main stem cortex (4), pith (5). (G-Os, LM x 75.)

1	Vascular gap
2	Stem vascular cylinder
3	Vascular system of bud
4	Stem cortex
5	Pith

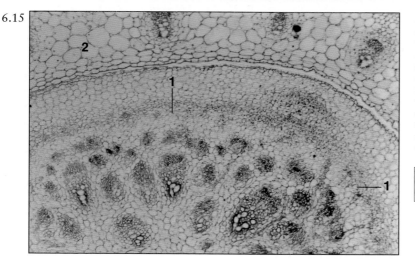

6.15   TS of the sub-apical region of *Zea mays* (maize) stem. Note the primary thickening meristem (1) in this monocotyledon (cf., **4.15**) from which numerous, scattered vascular bundles are derived internally, whilst peripheral parenchymatous derivatives lead to increase in stem thickness. Bundles frequently anastomose and their branches run obliquely outwards and upwards to supply the numerous veins of each leaf. Sheath of leaf primordium (2). (LM x 80.)

1	Primary thickening meristem
2	Leaf sheath

6.16 TS of the aquatic stem of the mono-cotyledon *Potamogeton* (pondweed). Extensive air chambers (1) occur in the cortex while the vascular tissue is confined to a narrow central cylinder demarcated externally by an endodermis (2). Eight variably-sized vascular bundles are present, with the xylem of each represented by a protoxylem lacuna (3). Pith (4), phloem (5). (LM x 75.)

6.16

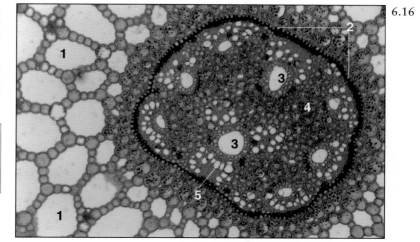

1	Air chambers
2	Endodermis
3	Protoxylem lacuna
4	Pith
5	Phloem

6.17 TS illustrating detail of the vascular tissue of the monocotyledon *Potamogeton* (pondweed) stem. This shows a prominent protoxylem lacuna (1) with wide sieve tubes (2) on either side (cf., **6.16**). Endodermis (3), fibres (4). (LM x 295.)

6.17

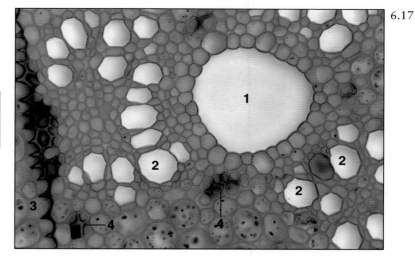

1	Protoxylem lacuna
2	Sieve tubes
3	Endodermis
4	Fibres

6.18 TS of a vascular bundle from *Phaseolus vulgaris* (bean) stem. Note the wide cambial zone (1) in which the cells are generally tangentially flattened; their radial alignment is abruptly distorted by the wider-diametered secondary xylem elements (2). The phloem is demarcated externally by the fibre cap (3) and, adjacent to the vascular cambium, secondary elements (4) are differentiating. (G-Os, LM x 285.)

6.18

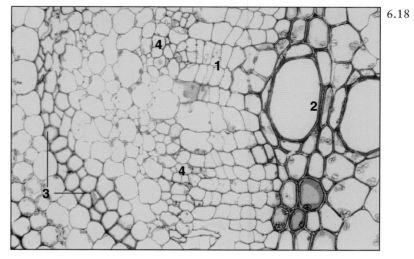

1	Cambial zone
2	Secondary xylem
3	Fibre cap
4	Secondary phloem

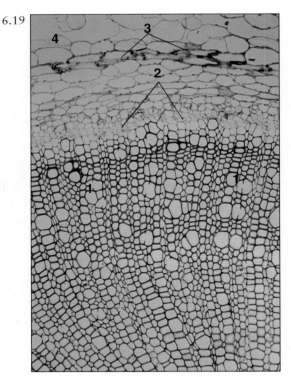

6.19 TS of the mature herbaceous stem of the dicotyledon *Linum usitatissimum* (flax). This dicotyledon has undergone a considerable amount of thickening during its one growing season. Note the radially-arranged secondary tracheary elements (1), arising from the vascular cambium (2). The phloem fibres (3) indicate the original position of the protophloem (cf., **1.4**). Cortex (4). (G-Os, LM x 115.)

1	Secondary xylem
2	Vascular cambium
3	Phloem fibres
4	Cortex

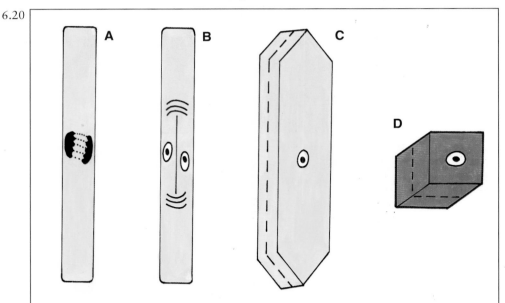

6.20 Diagrams of the vascular cambium shown in longitudinal views. An isolated fusiform initial is shown in radial view undergoing mitosis (**A**) and cytokinesis (**B**), while in **C** the two daughter cells are viewed from an oblique tangential aspect. **D** shows a ray initial which has just completed a tangential division.

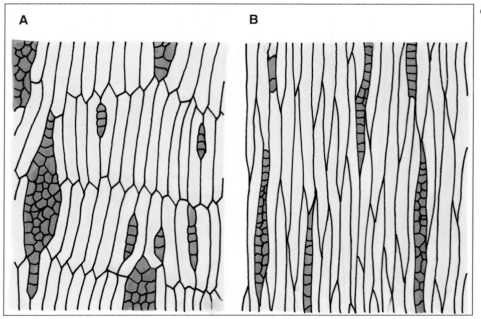

6.21 **A** and **B** show tangential views of a storied and non-storied cambium respectively (yellow indicates fusiform and red indicates ray initials).

6.22 Diagrammatic transverse section of the vascular cambium showing its pattern of tangential divisions. In **A**, division has occurred in the cambial mother cell (yellow) which leads in **B** to the formation of a phloem initial (blue). Meanwhile, the cambial cell is again dividing and gives rise in **C** to a xylem initial (red). The phloem initial in **C** is dividing to form two further initials and a new division (**D**) is already taking place in the outermost derivative. In the meantime, the cambial cell has formed a further xylem initial towards the inside (**D**), while the original initial has also divided, so that there are now three xylem initials.

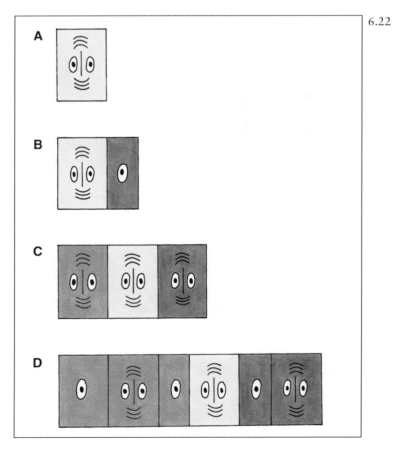

6.23

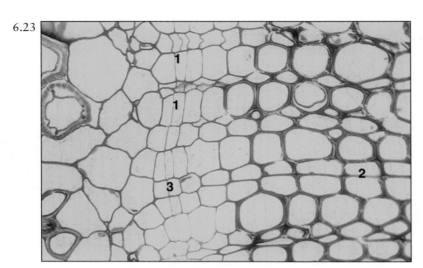

6.23 TS of the stem of the dicotyledon *Linum usitatissimum* (flax) showing detail of vascular differentiation. The tangentially-flattened cells of the cambial zone are well defined and, from its inner face, secondary tracheary elements are differentiating (cf., **6.19**). These undergo radial expansion but little tangential growth so that their origin from specific fusiform initials (1) can be traced. Radial rows of narrower thick-walled, ray parenchyma cells also occur (2); these originate from small ray initials (3). The phloem region is very narrow and demarcated externally by phloem fibres (cf., **1.4**). In the phloem the inner conducting elements are probably secondary but their derivation from the vascular cambium is obscure. (G-Os, LM x 300.)

1	Fusiform initials
2	Ray parenchyma cells
3	Ray initials

6.24

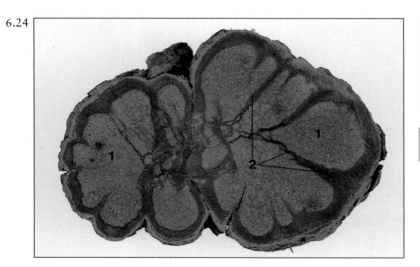

6.24 TS of the woody liane of the dicotyledon *Bauhinia*. In common with many other lianes, the stem undergoes anomalous secondary thickening. The originally regular xylem becomes split into highly lobed units (1) by the proliferation of the vascular parenchyma. Phloem wedges (2) extend between the xylem lobes.

1	Lobed xylem
2	Phloem wedges

6.25

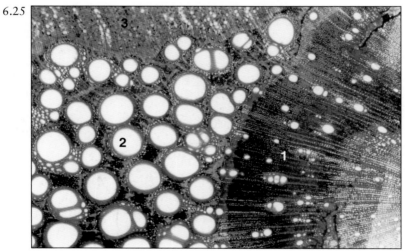

6.25 TS of the secondary xylem of a young stem of the dicotyledon *Bauhinia*. Note the radial regularity of the early secondary xylem (1) in contrast to the irregularity of the later wood which contains very numerous large lumened vessels (2). Vascular parenchyma wedges (3). (LM x 30.)

1	Secondary xylem
2	Large vessels
3	Vascular parenchyma wedges

6.26 TS of the stem of the liane *Tino-spora*. Note that the secondary xylem is deeply fissured by the proliferation of the intervening ray parenchyma (1). The xylem, in common with other lianes (cf., **6.25**) contains many large-lumened vessels (2) embedded in a sclerenchymatous ground tissue. (LM x 30.)

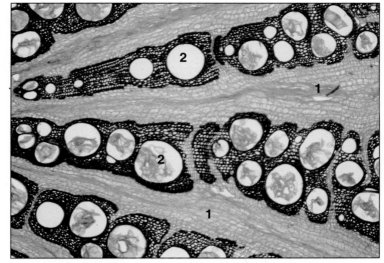

1	Ray parenchyma
2	Large vessels

6.27

6.28

6.27 The trunk of the conifer *Sequoiadendron giganteum* (giant redwood). These long-lived trees may grow up to nearly 80 metres in height and are protected by a very thick layer of soft cork. This sloughs off unevenly so that a number of growth layers are often visible.

6.28 Trunk of *Quercus suber* (cork oak) stripped recently of cork. The cork can be peeled away from the trunk of this dicotyledon and harvested (cf., **3.13**), leaving a thin layer of new cork still covering the functional secondary phloem on the trunk. The cork layer on the tree gradually builds up anew and will be harvested in about another 10 years.

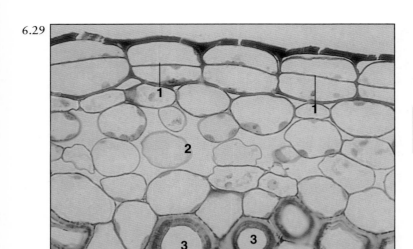

6.29

6.29 TS showing the epidermal origin of the cork cambium in the stem of the dicotyledon *Linum usitatissimum* (flax). Note the thin tangential walls (1) which divide the originally single-layered epidermis. Cortex (2), phloem fibres (3). (G-Os, LM x 300.)

1	Tangential division walls
2	Cortex
3	Phloem fibres

6.30

6.31

6.31 Outer surface of the trunk of a large *Cordyline australis* tree. This tree (cf., **1.22**) is a monocotyledon with anomalous secondary thickening; note its well-developed layer of protective cork.

6.30 Large tree of the dicotyledon *Acer pseudoplatanus* (sycamore) with a hollow trunk. Note how an extensive protective layer of periderm has grown over the exposed margins of the trunk and side branch.

6.32 TS showing distribution of lenticels on a young twig of the dicotyledon *Sambucus nigra* (elder). As secondary thickening proceeds the epidermis is replaced by cork in which numerous lenticels occur. Pith (1), primary xylem (2), secondary xylem (3). (LM x 8.)

1	Pith
2	Primary xylem
3	Secondary xylem

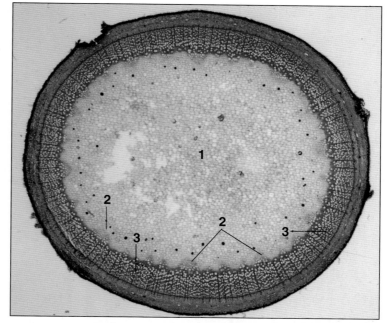

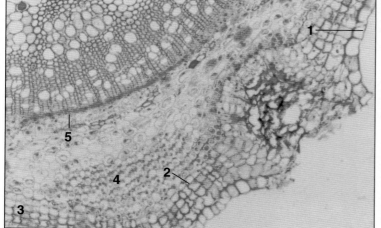

6.33 TS showing detail of a lenticel from the dicotyledon *Sambucus nigra* (elder) stem. The presence of the epidermis (1) indicates that the cork cambium (2) originated in the hypodermis. The cork cells (3) show radial alignment with cambial initials and to the inside a single layer of secondary cortex has differentiated. Collenchyma (4), vascular cambium (5). (LM x 85.)

1	Epidermis	4	Collenchyma
2	Cork cambium	5	Vascular cambium
3	Cork cells		

6.34 Large specimens of the bamboo *Dendrocalamus giganteus*. Note the uniform thickness of the trunks of these monocotyledons and the horizontal scars where the leaf sheaths were attached to the stem.

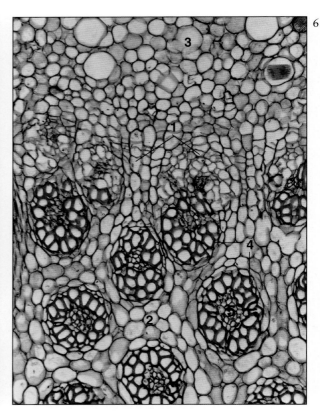

6.35　Large specimen of the arborescent monocotyledon *Dracaena draco* (dragon's blood tree). The tips of the branches bear crowded sword-shaped leaves and clusters of flowers. This several hundred-year-old specimen is endemic to the Canary Islands; its much-branched trunk has undergone extensive anomalous secondary thickening and is several metres wide at its base (cf., **6.36**).

6.36　TS showing detail of anomalous secondary thickening in *Dracaena* stem. The secondary thickening meristem (1) is cutting off discrete amphivasal vascular bundles centripetally and these are embedded in secondary parenchyma (2) which later becomes lignified. Cortex (3), xylem (4), phloem (5). (G-Os, LM x 145.)

1	Secondary meristem	4	Xylem
2	Secondary parenchyma	5	Phloem
3	Cortex		

CHAPTER 7

The root

Introduction

The basic functions of the root system are anchorage (**7.1** to **7.3**) and uptake of water and mineral nutrients (**1.12**). Roots are also concerned with the supply of cytokinins and gibberellins to the shoot system and commonly store starch (**1.14**). They sometimes form perennating tubers and may be modified in various other ways such as into contractile roots, pneumatophores (**7.4, 7.5**), prop roots (**1.26, 7.6**), aerial absorptive roots (**4.34**), haustoria (**7.7**), spines (**7.8**) and nitrogen-fixing nodules (**7.9, 7.10**). Roots also sometimes give rise to adventitious shoots (**3.21**).

In many dicotyledons the radicle of the embryo (**1.33**) develops into the vertically growing tap root; from this numerous lateral roots originate (**1.16, 7.9**). In monocotyledons the radicle is rarely persistent, but instead a fibrous root system develops from the base of the radicle while adventitious roots arise on the lower stem (**7.11**). Roots can penetrate several metres deep into the soil; the root systems of individual plants are sometimes very extensive and it has been calculated that a single mature plant of rye (*Secale*) may form a fibrous root system up to 40 times greater in surface area than the shoot system.

In trees and shrubs the absorptive roots are usually superficially located in the soil but the root system often extends laterally beyond their aerial canopies (**7.1**). Natural grafting between roots on the same tree (**7.2, 7.12**) is common and also frequently occurs between roots on different individuals of the same species. Grafting involves union of cambia and vascular tissue and this provides a possible route of disease transmission from an infected tree to others which are uninfected.

Anatomy of the mature primary root

In dicotyledons the core of the root is normally occupied by xylem, with commonly three to five protoxylem poles arranged around the central metaxylem (**4.31, 7.13**). However, roots with two poles (diarch roots) or more numerous poles (polyarch roots) also occur and sometimes a parenchymatous pith is present (**7.13**). Monocotyledonous roots are polyarch with numerous protoxylem poles (**1.29, 3.4**) and in some palms up to 100 poles occur. A parenchymatous or sclerenchymatous pith may be present (**4.34, 3.4**). In species with aerial prop roots a much wider pith is often present than in the underground root. In young monocotyledonous roots the xylem and phloem are separated by parenchyma but in older polyarch roots this may become sclerified (**3.4**), rendering it difficult to distinguish the boundary of the xylem.

Lateral and adventitious roots

Lateral roots develop endogenously (**4.9**) from the pericyclic parenchyma. They are frequently initiated in the parent root at about the level of the oldest root hairs (**4.27**). In dicotyledons they arise opposite a protoxylem pole (**4.9**) so that the xylem linkage between lateral and parent root is direct. In monocotyledons the root primordia often originate between the xylem poles.

In the initiation of a lateral root a number of pericyclic parenchyma cells are involved which undergo division in varied planes and dedifferentiate to a meristematic state (**7.14**). In *Zea* the thickened, partly lignified, pericyclic parenchyma walls undergo thinning before the cells divide. Eventually a dome of meristematic cells is formed which begins to protrude into the root cortex (**4.9**). The lateral root primordium is invested by the endodermis of the parent roots. This initially undergoes anticlinal divisions to accommodate the growth of the lateral primordium but the endodermis disintegrates later. Before the lateral root emerges it penetrates the overlying cortex and epidermis of the parent root. A well-defined root cap is formed and internally the apical organisation characteristic of the species established (**7.15**). A quiescent centre is initially absent from the root primordium but this develops soon after the emergence of the young root.

Adventitious roots are often present on the stem (**1.26, 3.18, 7.8, 7.11**), frequently located at the nodes (**4.18**). They usually have an endogenous origin, either from the ray parenchyma or vascular parenchyma (**4.18**). Roots may also develop in conjunction with adventitious buds on excised leaves (**4.27, 7.16**).

In a number of species of tropical and subtropical origin an adventitious system of prop roots is present (**1.26, 7.6**). In the aptly-named strangler fig (*Ficus*, **7.12, 7.17**) the root system develops from a bird-transported seed germinating in the canopy of the host tree. The roots grow downwards on the trunk, fuse around it and eventually kill the host tree, whose trunk rots away to leave a hollow fig tree (**7.17**). In tree ferns a dense, tangled mat of adventitious roots invests the trunk and provides additional mechanical support to the plant (**7.18**).

Mycorrhizae and root nodules

The root systems of most plants are associated with fungi (mycorrhizae, **7.19**) in a symbiotic relationship. The fungus supplies the plant with inorganic nutrients, especially phosphorus, while the fungus receives sugars and other organic nutrients from the plant. A minority of plants (*Betula*, *Fagus*, *Pinus*) are associated with ectomicorrhizae in which the fungal hyphae form a dense covering over the roots (**7.19**) and also ramify within the intercellular spaces of the root cortex. The roots of most species, however, show endomicorrhizae: the fungus is situated within the apoplastic system of the root cortex and, where the hyphae penetrate the cortical cell walls, they remain separated from the root protoplast by the plasmalemmae.

The nodules in leguminous roots (**7.9**) are formed in a symbiotic relationship with the bacterium *Rhizobium*, but in many non-leguminous nitrogen-fixing plants (e.g. *Alnus*, *Comptomia*) the root nodules (**7.10**) develop in association with the actinomycete *Frankia*. The lobes of the nodules represent modified root tips (**7.20**) which sometimes grow out into normal roots (**4.29**). The vesicular stage of *Frankia* in the root nodules (**7.21**) is able to fix free nitrogen in the soil atmosphere; thus the plant receives nitrogenous compounds while the micro-organism obtain sugars in return.

Secondary growth in roots

In the majority of herbaceous dicotyledons some secondary growth occurs (**1.25, 7.22**). It is extensive in woody gymnosperms and angiosperms (**7.1, 7.2, 7.17, 7.23 to 7.25**), and occurs in the arborescent monocotyledon *Dracaena*. In the former, the vascular cambium is initiated from parenchyma cells which resume division and first become active between the metaphloem and the metaxylem (**4.31, 7.25**). These divisions progressively spread laterally over the protoxylem poles to involve the pericyclic cells, so that the vascular cambium completely encloses the xylem core (**1.25, 7.25**) and eventually becomes circular (**7.22 to 7.25**). In the secondary xylem wide rays may occur, especially on the radii of the protoxylem poles, but narrower rays are more generally distributed (**7.23, 7.24**).

The secondary xylem of angiosperm roots contains large-diametered vessels which are often interspersed between smaller tracheary elements (**7.22, 7.24**), but in gymnosperms the absence of vessels provides a more uniform appearance to the xylem (**7.23**). The formation of secondary vascular tissue causes an expansion of the core of the root and generally the endodermis and external tissues are sloughed off (**7.25**). However, before this occurs a phellogen arises endogenously within the proliferated pericyclic parenchyma. The vascular cylinder becomes invested and protected by a layer of cork (**7.24**) while growth rings in the xylem may also become apparent (**7.25**). In some herbaceous dicotyledons the cork cambium arises superficially (**7.22**). In aborescent species with thickened roots (**7.1, 7.2**), their radial expansion can exert great pressure and help crack open rocks (**7.26**).

In *Daucas* (carrot, **7.27**) and other fleshy storage roots the secondary vascular tissue mainly differentiates as parenchyma rather than conducting elements. In *Ipomoea* and *Beta* the original vascular cambium is replaced by further cambia which form successively deeper in the extensive parenchymatous tissue of the secondary xylem (**7.28**).

7.1 Root system of a mature tree of the dicotyledon *Fraxinus excelsior* (ash). This was growing on a steep, eroded slope and shows the richly-branched, superficial roots which extend laterally some metres from the trunk.

7.2 Detail of the superficial root system of the dicotyledon *Fagus sylvatica* (beech). This was a large tree and the roots growing in the lower side of a steep slope showed frequent grafting; this bracing provides additional stability for the trunk and aerial canopy.

7.3 Buttress root of a large rain forest tree. This unidentified specimen (growing in Queensland, Australia) had several prominent buttresses at the base of its trunk which extended laterally several metres and were a metre or so high at their origin from the trunk. Buttresses occur on a number of rain forest tree species and may reach up to several metres high. They help to stabilise large trunks and their aerial canopies; the buttresses have numerous lenticels and are also thought to aid in aeration of the underground root system.

7.4

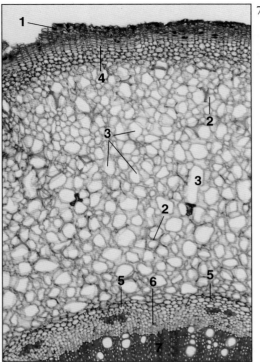

7.5

7.4 Pneumatophores of *Avicennia nitida* (mangrove) growing in saline estuarine mud. The negatively geotropic roots of this dicotyledon develop from long horizontal roots which extend from the base of the tree trunk and grow in the oxygen-depleted mud. The breathing roots are protected by a cork covering but contain many lenticels to allow aeration of the horizontal root system.

7.5 TS of a small pneumatophore of the dicotyledon *Avicennia nitida* (mangrove). This negatively geotropic root (cf., **7.4**) is covered by cork (1) containing numerous lenticels. The numerous cortical sclereids (2) support the parenchyma cells which are separated by extensive intercellular spaces (3) providing aeration to the submerged horizontal roots. Cork cambium (4), endodermis (5), secondary phloem (6), secondary xylem (7). (LM x 70.)

1	Cork	5	Endodermis
2	Cortical sclereids	6	Secondary phloem
3	Intercellular spaces	7	Secondary xylem
4	Cork cambium		

7.6

7.6 *Rhizophora mangle* (spider mangrove) growing in saline estuarine mud. These dicotyledons show a dense tangle of adventitious stilt roots which help to stabilise the tree and transport water and nutrients to its trunk. Although covered by bark, numerous lenticels in the stilts allow aeration of the roots growing in the oxygen-depleted mud. In the foreground several seedlings, which germinated in the fruits while still attached to the tree, have fallen into the mud and started to grow.

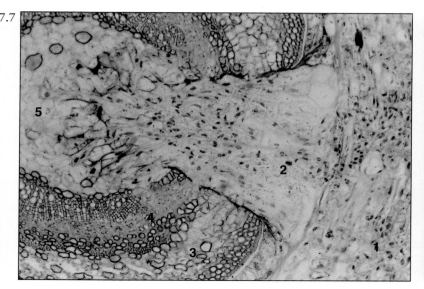

7.7 Haustorium of the parasitic dicotyledon *Cuscuta* (dodder). The leafless, non-green stem (1) of dodder loses its seedling root system. However, as the dodder twines itself anticlockwise around the host nettle or gorse stem, it forms a haustorium (2) at each point of contact. This highly modified root penetrates the host epidermis and cortex (3), to tap the vascular tissue (4) and pith (5) for water and nutrients. (LM x 75.)

7.8 Long spines on the trunk of the palm *Chorisia speciosa* (a monocotyledon). These modified adventitious roots are of limited growth and the apices and root caps are replaced by sclerenchymatous points.

1	Stem	4	Vascular tissue
2	Haustorium	5	Pith
3	Cortex		

7.9 Root system of the dicotyledon *Glycine soja* (soybean) showing nodules. This plant was grown in water culture containing balanced mineral salts, except for the absence of nitrates or ammonium salts. Nevertheless, the symbiotic bacterium *Rhizobium* present in the root nodules was able to fix free nitrogen so that the soybean plant grew vigorously. Legumes commonly develop root nodules which, on their decay, enrich depleted soils.

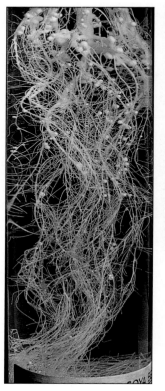

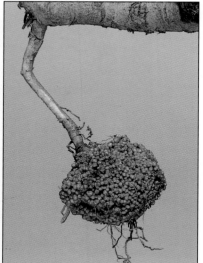

7.10 Large compound nodule in the root of the dicotyledon *Alnus glutinosa* (alder). This is composed of numerous closely crowded individual nodules which are infected with the symbiotic actinomycete *Frankia*. which is able to fix free nitrogen in the soil atmosphere.

7.11   Adventitious roots at the base of a palm trunk. This unidentified specimen was growing in the rain forest in Queensland, Australia.

7.12   Roots of strangler fig (*Ficus microcarpa*, a dicotyledon) investing a rain forest tree. These roots are growing down its trunk and have developed from a seed germinating in the droppings of a bird in the canopy of the 'host' tree. The woody roots frequently graft together and, as the 'host' tree trunk expands, eventually cut off the food supply so that it dies (cf., **7.17**).

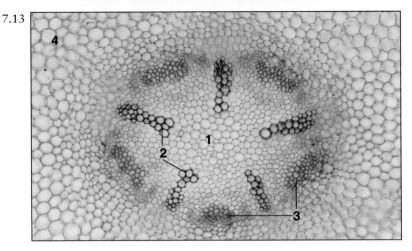

7.13   TS of the primary root of the dicotyledon *Vicia faba* (broad bean). Note the extensive parenchymatous pith (1), five arcs of xylem with the metaxylem (2) situated centripetally, and the strands of phloem (3) alternating with the xylem. The endodermis is not yet clearly defined and no vascular cambium is apparent. Cortex (4). (LM x 85.)

1	Parenchymatous pith
2	Metaxylem
3	Phloem
4	Cortex

7.14 LS of a lateral root primordium of the dicotyledon *Pisum sativum* (pea). This has formed by the division and de-differentiation of the pericyclic parenchyma of the parent root. The densely cytoplasmic cells of the primordium (1) bulge out into the parenchyma of the parent root cortex (2). (G-Os, LM x 285.)

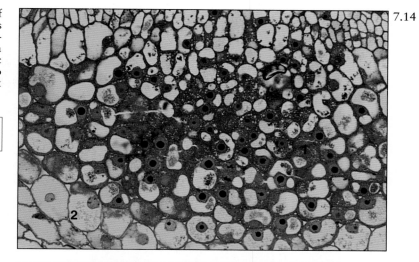

7.14

1 Root primordium
2 Root cortex

7.15 LS of lateral root primordia in the dicotyledon *Salix* (willow). These originated opposite protoxylem poles from the pericyclic parenchyma of the parent root (cf., **4.9**) and have now penetrated its cortex (1), with the root tips distorting the normally circular epidermis (2). The vascular cylinder (3) of one root primordium links with the vascular system of the parent root. Root apical meristem (4), root cap (5). (LM x 200.)

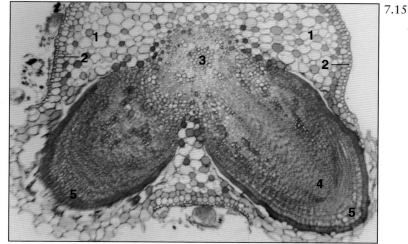

7.15

1 Cortex
2 Epidermis
3 Vascular cylinder
4 Root apical meristem
5 Root cap

7.16 *De novo* root and bud production on leaf cuttings of the monocotyledon *Sansevieria trifasciata* (mother-in-law's tongue). The basal end of each was inserted in moist compost and all show adventitious roots at the basal ends. Two cuttings have also developed large leafy buds connected by short rhizomes to the parent leaf segments.

7.16

7.17

7.18 TS of the old trunk of *Cyathea* (tree fern). Note in this rain forest fern the thick mantle of adventitious fibrous roots (1) which arise from and invest the petiolar bases (2). The latter persist after the fronds wither and surround a complex sclerenchymatous vascular core (3), but secondary growth is absent. The leaf bases and mantle of roots help support the stem which may be several metres high and bears evergreen fronds up to 4 metres long (cf., **1.7**).

1	Fibrous roots	3	Vascular core
2	Petiolar bases		

7.17 Hollow 'trunk' of a strangler fig tree (*Ficus microcarpa*). This actually represents the fused roots of an epiphytic seedling germinating in the crown of a rain forest tree and growing down its trunk to the soil. Meanwhile the 'host' tree's vascular system has been cut off by the investing roots (cf., **7.12**) of the fig, and the tree trunk eventually dies and rots away leaving the 'trunk' of the fig in its place.

7.19

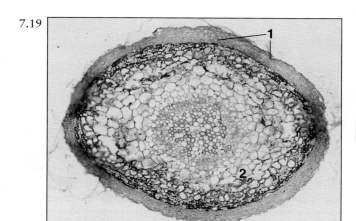

7.19 TS of a mycorrhizal root of the dicotyledon *Fagus sylvatica* (beech). Note the investing mat of symbiotic fungal mycelium (1); the hyphae penetrate into the surrounding soil and absorb phosphates to supply to the root. They also penetrate between the root epidermal cells into the cortex (2), but their distribution remains apoplastic. (LM x 90.)

1	Fungal mycelium	2	Root cortex

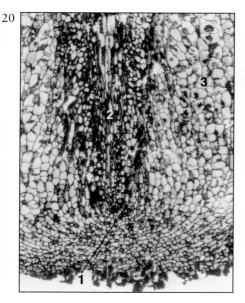

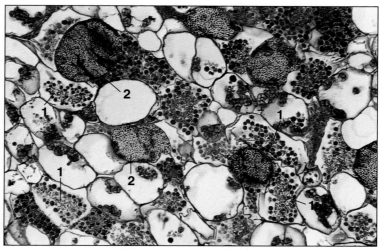

7.21   Cortical tissue of *Alnus glutinosa* (alder) root nodule showing the distribution of the symbiotic actinomycete *Frankia*. This section shows host cells containing numerous endophytic stages of the actinomycete. Note in many cells (1) the fine hyphal network bear swollen vesicles (which fix nitrogen), while other cells (2) contain the sporangia. (G-Os, LM x 285.)

1	Hyphae with swollen vesicles
2	Host cells containing sporangia

7.20   LS of a root nodule of the dicotyledon *Alnus glutinosa* (alder). This single nodule was part of a much larger compound nodule (cf. **7.10**) and represents a swollen lateral root of determinate growth. Note the apical meristem (1), vascular cylinder (2) and cortex (3). No *Frankia* is evident in this section but the cortical tissue basal and lateral to this area showed early stages of infection (cf., **7.21**). (LM x 100.)

1	Root apical meristem
2	Vascular cylinder
3	Cortex

7.22   TS showing early secondary growth in the fleshy root of the dicotyledon *Ipomoea batatas* (sweet potato). In the centre a cylinder of secondary xylem with prominent vessels (1), invests the core of primary xylem (2). At a later stage the vascular cambium (3) produces large quantities of xylem parenchyma leading to the formation of a root tuber. Note the cortical origin of the cork cambium (4). (LM x 75.)

1	Secondary xylem	3	Vascular cambium
2	Primary xylem	4	Cork cambium

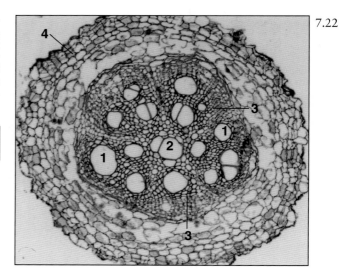

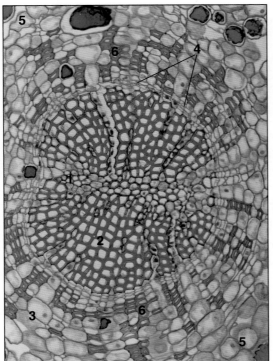

7.23 TS showing early secondary thickening of the young root of the gymnosperm *Ginkgo biloba*. The primary xylem is diarch and the narrower protoxylem elements (1) contrast with the somewhat wider but uniform tracheids of the secondary xylem (2). Both this and the secondary phloem (3) show radial cell lineages traceable to their origin in the vascular cambium (4). Cortex (5), phloem fibres (6). (LM x 130.)

1	Protoxylem elements	4	Vascular cambium
2	Secondary xylem	5	Cortex
3	Secondary phloem	6	Phloem fibres

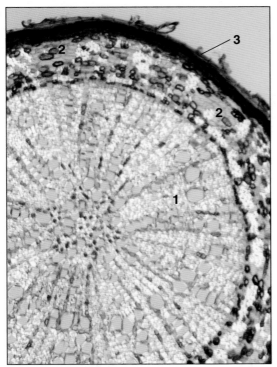

7.24 TS of the young secondary-thickened root of the dicotyledon *Tilia cordata* (lime). A broad cylinder of secondary xylem (1, with many wide vessels) is surrounded by a narrow layer of secondary phloem (2) which contains numerous fibres. A thick layer of cork (3) covers the root surface. (Polarised LM x 75.)

1	Secondary xylem	3	Cork
2	Secondary phloem		

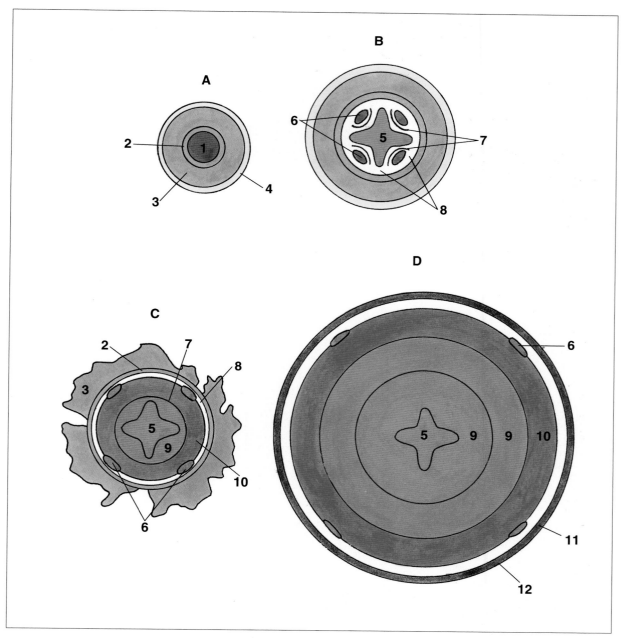

7.25 Diagrammatic development of the root in a woody dicotyledon as seen in TS. **A,** just behind the root apex a central cylinder of procambium (1), is separated by the single-layered endodermis (2) from the cortex (3) and epidermis (4). **B,** beyond the root hair zone the mature primary root shows a tetrarch xylem (5) separated from the phloem strands (6) by narrow arcs of vascular cambium. (7). **C,** the vascular cambium has differentiated laterally from the pericycle (8) and now encloses the xylem. The vascular cambium has given rise to a layer of secondary xylem (9) internally and secondary phloem (10) externally. Primary phloem strands are visible on its outer margin while the primary xylem remains intact internally. The epidermis has ruptured and the cortex and endodermis are being sloughed off. **D,** the root shows further secondary-thickening with two growth rings in the secondary xylem. These are not evident in the secondary phloem although the crushed remains of the primary phloem are still evident. The endodermis and external tissues have sloughed off and a cork cambium (11) has arisen in the pericycle to form a protective layer of cork (12).

1	Procambium
2	Endodermis
3	Cortex
4	Epidermis
5	Tetrarch primary xylem
6	Primary phloem strands
7	Vascular cambium
8	Pericycle
9	Secondary xylem
10	Secondary phloem
11	Cork cambium
12	Cork layer

7.26 Fine roots from a large tree of the dicotyledon *Fraxinus excelsior* (ash) growing in the cracks of a rock face. These roots have grown down from the main branch roots in the soil above and their expansion by secondary thickening growth, helps to further break open the exposed rock face.

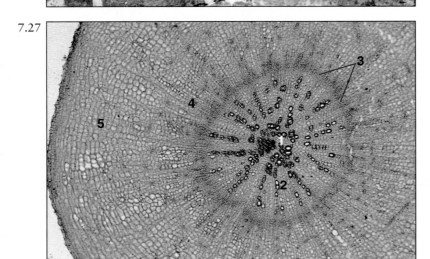

7.27 TS of the fleshy root of the dicotyledon *Daucus carota* (carrot). The narrow core of primary xylem (1) is surrounded by an extensive but mainly parenchymatous secondary xylem (2) in which a few tracheary elements occur. The well-defined vascular cambium (3) also produces centrifugally a largely parenchymatous secondary phloem (4). Cortex (5). (LM x 30.)

1	Primary xylem
2	Secondary xylem
3	Vascular cambium
4	Secondary phloem
5	Cortex

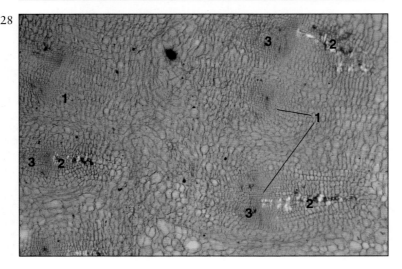

7.28 TS showing detail of the fleshy root of the dicotyledon *Beta vulgaris* (beetroot). The vascular cambium produces extensive parenchyma in the secondary xylem; supernumery cambia (1) arise within this tissue giving rise to discrete strands of xylem (2) and phloem (3). (LM x 30.)

1	Supernumery cambia
2	Xylem strands
3	Phloem strands

144

CHAPTER 8

Sexual reproduction

Introduction

The great majority of ferns are homosporous, with numerous haploid spores being produced by meiosis in the sporangia borne on the leaves of the diploid plant (sporophyte, **1.7**). When the spores are shed they develop into small, but free-living, green gametophytes (**8.1**) on which the gametes are later formed. Fusion of the motile sperm with the egg results in a diploid embryo; this re-establishes the sporophyte generation which is soon independent of the gametophyte.

The seed plants are heterosporous and do not show a free-living gametophyte generation. In the gymnosperms a single haploid megaspore matures to form the embryo sac within the naked female ovule (**1.5**, **1.32**). This megaspore is formed by the meiosis of a solitary megaspore mother cell located in the diploid nucellus (megasporangium). Only one of the four resultant spores matures while the others abort (**1.32**). Both the embryo sac and nucellus are enclosed within an integument and these structures collectively constitute the ovule (**1.32**) which is retained on the tree (**1.5**). The haploid megaspore divides repeatedly to form a mass of gametophytic tissue within the enlarged embryo sac, and eventually one or more eggs become demarcated within this tissue close to the micropyle (**8.2**).

Following pollination of the ovule by a microspore, the egg is fertilised by a sperm and the diploid embryo develops. The testa of the mature seed is formed from the modified integument of the ovule, while copious food reserves for the future seedling are contained within the tissue of the female gametophyte surrounding the embryo (**1.32**, **8.2**).

In flowering plants the ovule is enclosed within an ovary (**1.32**) in contrast to the exposed ovule in gymnosperms (**1.5**). The female gametophyte within the embryo sac is typically reduced to eight cells and one of these, at the micropylar end of the embryo sac, represents the egg (**1.32**). In contrast to gymnosperms double fertilisation occurs in angiosperms with one sperm fertilising the egg to form the diploid embryo (**1.33**) while the other sperm fertilises the two central polar nuclei to form the triploid, nutritive endosperm (**1.32**) which is initially coenocytic.

General features of flowers

Most flowers contain both male and female sex organs (**1.1**, **1.31**) but some flowers are unisexual (**8.3**) and may occur on the same plant (**8.3**) or separate individuals. The flower of *Magnolia* (**1.31**) illustrates the general features of a dicotyledon. The upper half of the elongate receptacle (floral axis) is covered by a large number of spirally arranged and separate female organs (carpels) which collectively constitute the apocarpous gynoecium. Below the carpels numerous spirally arranged male organs (stamens constituting the androecium) are inserted onto the receptacle by short filaments. These are terminated by large anthers containing the pollen (microspores). At the base of the receptacle there are usually nine large, petaloid, perianth members.

In contrast with *Magnolia*, the monocotyledonous flower of *Lilium* (**1.1**) has six petaloid perianth members arranged in two alternating whorls. These encircle six stamens, while the syncarpous gynoecium consists of three fused carpels. The three compartments of the ovary are clearly visible on the fruits developing after the flowers wither (**1.1**). The ovary in *Lilium* is terminated by a long, slender style which is tipped by a slightly swollen stigma. As in *Magnolia* the ovary is superior since its point of insertion lies above the rest of the floral parts, but in other flowers the ovary may be inferior (**8.4**). In the early development of most flowers the organ primordia arise in centripetal sequence so that the perianth is initiated first and the gynoecium last.

Perianth

In some flowers the perianth members are similar in appearance (**1.1**, **1.31**) but in many others the perianth is differentiated into the outer sepals, constituting the calyx, and the petals forming the corolla (**8.6**, **8.7**). There are also some flowers (usually wind-pollinated) in which the perianth is vestigial (**3.30**, **8.8**).

The calyx consists of several sepals that are frequently green and leaf-like (**1.6**, **5.7**) but their mesophyll is not usually differentiated into palisade and spongy layers. Besides their protective role in investing the unopened flower bud, sepals are sometimes brightly coloured (**8.6**) and attractive to pollinators; while in *Taraxacum* the hairy pappus (calyx, **8.9**) of each floret assists in the wind dispersal of the numerous small fruits (**8.10**).

In most flowers the corolla consists of a number of petals (**8.11**) which are variously modified to attract pollinators (**5.7, 8.6, 8.7**) but the corolla is greatly reduced or absent in flowers where the pollen is wind-distributed (**8.8**). In animal-pollinated flowers nectaries are often present (**3.30, 8.6**) and their sugary secretion attracts pollinators; in many bee-pollinated flowers the epidermal cells of the petals contain ultraviolet-absorbing flavonoid pigments which are visible to the bee and act as guides to the nectaries (**8.12**). Betacyanins in the epidermal vacuoles of the petals and carotenoids in the chromoplasts often cause bright colouring (**8.6, 8.7**) which attracts pollinators.

Petals are usually ephemeral structures with a thin cuticle and few stomata, while the mesophyll is usually non-photosynthetic. In actinomorphic (radially symetrical) flowers perianth members are all of similar size and distributed regularly around the receptacle (**1.1, 8.6, 8.11, 8.12**). However, many flowers are bilaterally symmetrical (zygomorphic, **5.7, 8.7**). As with the other floral organs, the petals are often fused (**8.5, 8.13**) although sometimes their tips remain free (**5.7**).

Androecium

Each stamen is terminated by an anther (**1.1, 1.31**) which contains the pollen grains (**8.14, 8.16**). The anther is borne on a filament which is either inserted directly onto the receptacle or the corolla. The filaments are sometimes of unequal length (**8.15**), and may be very short (**1.31**) whilst in orchids filaments are absent. In some flowers the filaments are fused to form a tube round the gynoecium and in the Compositae the anthers are fused (**8.13**). A vascular bundle runs along the filament (**8.17**) and supplies water and nutrients to the anther. The anther is commonly bilobed (**8.15**) with two cylindrical pollen sacs present in each lobe (**8.13, 8.18**). The ripe anthers usually split longitudinally to release their pollen (**8.14, 8.18**), but in some plants (e.g. *Rhododendron*, heathers) the pollen is released from apical pores (**8.16**).

In the young anther (**8.5**) several layers of hypodermal cells divide periclinally within each pollen sac to give rise to a central mass of sporogenous cells and the investing parietal tissue (**8.13**). The latter differentiates into the hypodermal endothecium, an intermediate parenchymatous layer and the tapetum which surrounds the sporogenous tissue (**8.19**). The endothecial walls normally develop numerous bands of thickenings on the anticlinal and inner periclinal walls (**8.20**) which may be lignified. Most anthers dehisce along specialised longitudinal tracts of epidermal cells (stomia, **8.18, 8.20**) which are underlain by unthickened endothecial cells. In species such as *Phaseolus vulgaris* (bean), self pollination occurs in the unopened flower (cleistogamy). The grains germinate *in situ* and the pollen tubes penetrate the anther wall to reach the stigma.

The tapetum plays a vital part in the nutrition of the developing sporogenous tissue, and consists of a layer of densely cytoplasmic cells (**8.19**) which are commonly bi- or multinucleate (**8.21**). In many anthers these cells remain intact but in species with an invasive (amoeboid) tapetum their walls degenerate (**8.22**). As a result a coenocytic periplasmodium is formed from their combined protoplasts, which directly invests the developing sporogenous cells. Within each anther the pollen mother cells typically undergo synchronous meiotic divisions (**8.21, 8.22**).

In many dicotyledons the four immature haploid pollen grains (microspores) formed by meiosis of the pollen mother cell (microsporogenesis) lie in a tetrahedron (**8.22**). Hence each grain has three inner faces in contact with the other grains. However, an isobilateral segmentation pattern is more frequent in monocotyledons. The pollen grain is initially uninucleate (**8.23**) but the first mitotic division, that often occurs in the anther, divides the maturing grain into a large vegetative and a smaller generative cell. In *Drosera* the pollen grains remain united as tetrads, but in the majority of flowering plants the four grains separate as the investing callosic wall of the pollen mother cell dissolves. In many orchids and *Rhododendron* the pollen adheres together in a sticky mass termed a pollinium (**8.16**).

The outer wall of the mature pollen grain (exine) is extremely complex (**8.23 to 8.25**) and comprises an inner layer of sporopollenin (nexine, which is very resistant to decay), and the outer sexine. The latter is permeated by sporopollenin and is composed of fused rods (baculae). The exine pattern of different species is very variable (**8.24, 8.25**) and of great taxonomic significance. The inner region of the pollen grain wall is cellulosic and is termed the intine.

Gynoecium

In many flowers (**8.11, 8.13, 8.14**) the carpels show fusion (syncarpy). In such flowers the ovaries are united but sometimes the styles remain separate (**8.14**) or the stigmas are lobed (**8.26**). While most floral parts are generally ephemeral the ovary, containing the ovules, continues growth after fertilisation and develops into the fruit (**1.6, 8.27**).

The stigma is the pollen receptor (**1.32, 8.11**); in wind-pollinated flowers the stigmas are often feather-like, while in other plants the stigmatic epidermis is either papillose or hairy (**8.26**). The majority of carpels possess a style (**8.13 to 8.15**) but in some species this is very short (**1.31**). In syncarpous species in which the styles are also fused, there is usually one main longitudinal vascular bundle per stylar component (**8.13, 8.28**). The centre of the style consists of transmitting tissue (**8.28**) and, after the pollen grains germinate (**8.29**), the pollen tubes grow downwards through this tissue towards the ovary (**1.32**), absorbing nutrients from this tissue *en route* (**8.30**).

The ovules are attached to a thickened region of the ovary wall termed the placenta (**1.32**). The carpel is generally interpreted as a folded and modified leaf with its abaxial surface outermost; its margins are normally fused and typically two longitudinal placentae lie internally (adaxially), close to the fused leaf margins (**8.31**). In syncarpous ovaries axile placentation is common, with the margins of the carpels fused at the centre of the ovary (**1.6, 8.27**). However, the placentae may also be located at the outer margin of the ovary (parietal placentation) or basally (**8.4**). In the apocarpous ovary marginal placentation is common (**8.31**).

Each ovule is connected at its base (chalaza) to the placenta via the funiculus (**1.32, 8.32, 8.33**). The ovule is invested by two (or sometimes one) thin integuments enclosing the nucellus. At the apical end of the ovule a narrow channel, the micropyle, penetrates the integuments (**1.32**) to expose the surface of the nucellus. The orientation of the ovule relative to the funiculus is variable but commonly the funiculus is bent through 180 degress (anatropous ovule, **1.32, 8.33**). The ovary walls and placentae are usually richly vasculated (**8.31**) and a number of smaller veins branch into the ovules (**8.33**) but vascular tissue rarely extends into the nucellus.

Early in development of the ovule a megaspore mother cell becomes demarcated at the micropylar end of the nucellus (megasporangium, **8.34**) and undergoes meiosis. In the commonest situation (*Polygonum*, monosporic type) a vertical row of four haploid megaspores is formed. Only the deepest-sited (chalazal) cell develops further while the outer three megaspores degenerate (**8.35**). The single functional megaspore enlarges greatly within the expanding nucellus and develops into the embryo sac (**1.32, 8.33, 8.35**). The original single haploid nucleus divides (**8.35**) to give the eight nuclei characteristic of the mature embryo sac (megagametophyte, **1.32**).

Other patterns of megaspore mother cell division also occur. In the bisporic type, one of the two derivatives of the first meiotic division undergoes further divisions and gives rise to the eight nuclei of the embryo sac while the other derivative degenerates. In tetrasporic development, the megaspore mother cell (**8.34**) undergoes meiosis but all four nuclei are confined within the common cytoplasm of the embryo sac. These nuclei then undergo a variable number of mitoses, so that the mature embryo sac often contains more than eight nuclei. In the semi-mature embryo sac of *Lilium* (**8.36**) one haploid nucleus is located at the micropylar pole but the other three nuclei at the chalazal end fuse to give a triploid nucleus. Subsequent mitoses and rearrangement of the nuclei lead to three haploid nuclei being situated at the micropylar end, one haploid plus one triploid nucleus in the centre of the embryo sac, and three triploid nuclei at the chalazal end.

In the common monosporic type the eight nuclei and associated cytoplasm become separated by thin walls and are located in various regions of the embryo sac (**1.32**). Three antipodal cells are located at the chalazal end of the sac, two synergids and a median egg cell lie at the micropylar end while the central region contains two polar nuclei within the endosperm mother cell. The egg cell is normally larger than the synergids; its nucleus and most cytoplasm usually lie towards the chalazal pole and here the wall is scantily developed. In the synergids, the cytoplasm is concentrated at their micropylar poles and their walls may be modified into a wall labyrinth (filiform apparatus) similar to that present in transfer cells (**2.63**). The synergids apparently transport nutrients from the nucellus to the egg and may form absorptive haustoria within this tissue. In some species the synergids have a role in fertilisation. The antipodal cells commonly degenerate before fertilisation of the egg, but they may persist and also develop haustoria.

Fertilisation

When a compatible pollen grain is deposited on the stigma (**1.32**) a pollen tube grows out through one of the germination pores in the pollen grain wall (**8.25**) where the exine is poorly developed. The cytoplasm and the nuclei migrate into the pollen tube which, after penetrating the stigma, grows downwards in the stylar transmission tissue (**8.30**).

In the anther, the generative nucleus of the pollen grain may already have divided into two sperm nuclei; otherwise this occurs within the pollen tube (**1.32**). The nuclei and cytoplasm concentrate at the tip of the elongating pollen tube; behind this the tube is highly vacuolate and is often sealed off by plugs of callose. The pollen tube frequently enters the embryo sac via the micropyle (**1.32**) but in rare cases may directly penetrate the integuments. Once entry to the embryo sac has been gained, a pore forms near the tip of the pollen tube and the two sperms are liberated. One sperm nucleus fuses with the egg nucleus and the other fertilises the centrally-located polar nuclei (**1.32**). In most flowering plants the cytoplasm surrounding the sperm nuclei is not transmitted at fertilisation and therefore inheritance of chloroplasts and other organelles is generally via the female line.

Development of the seed
Introduction
The fertilisation of the haploid egg by a haploid sperm gives rise to a diploid zygote which subsequently divides repeatedly and develops in a highly organised manner into the embryo (**1.33**). The mature embryo commonly undergoes a period of dormancy within the protective seed coat (testa) which develops from the integuments of the enlargened ovule (**1.33**). The embryo is packed with food reserves (**2.54**); in albuminous seeds further food is stored in the mass of endosperm surrounding the embryo. In the Caryophyllaceae little endosperm is formed but the nucellus develops into a nutritive perisperm.

Embryo development
Following entry of the sperm nucleus through the wall-free region of the egg cell, a wall is secreted in this area of the zygote. Its nucleus then undergoes mitosis and usually a transverse wall divides the zygote into basal and terminal cells. The basal cell divides mainly transversely to form a suspensor (**1.33**); this pushes the terminal cell away from the micropyle and into the endosperm which is developing (**1.33, 2.5, 8.37, 8.38**) within the expanding embryo sac. In legumes the suspensor cells often show highly polyploid, amoeboid, nuclei (**8.39**) and apparently synthesise growth substances of importance for the development of the embryo. The terminal cell undergoes divisions in various planes to form a globular proembryo (**8.38**) and soon an outer layer of anticlinally dividing protoderm cells becomes established. In dicotyledons the enlarging globular proembryo becomes transformed into a heart-shaped structure (**1.33**) as paired cotyledons at its chalazal end develop.

The embryo now elongates and differentiation of the procambium and ground tissues occurs (**8.40**). The radicle apex becomes demarcated at the micropylar pole of the embryo and merges into the hypocotyl above (**8.41**). Meanwhile, at the other end of the hypocotyl and between the cotyledons, a bulge representing the plumule becomes apparent. In the mature seed the plumular apex may either remain small (**8.41**) or is larger and has already given rise to its first foliage leaves (**4.1, 8.42, 8.43**), while the radicle shows a root cap and apex (**8.44**). Depending upon the architecture of the embryo sac the embryo may be straight or variously curved (**1.33, 8.41**). In albuminous seeds abundant endosperm is present and the cotyledons are thin and leaf-like, in contrast to their swollen appearance in non-endospermous seeds (**8.41, 8.42**).

In monocotyledons, the early development of the embryo parallels that in dicotyledons, but only a single lateral cotyledon is formed (**1.21**). Monocotyledonous seeds are commonly albuminous with the cotyledon apparently acting as a digestive organ. In palms the cotyledon often enlarges greatly and becomes haustorial, whilst in grasses the scutellum, which divides the endosperm from the embryo proper, is sometimes regarded as a modified cotyledon. During early germination the plumule of grasses is protected by a cylindrical leaf-like coleoptile while the radicle is initially ensheathed by the coleorhiza (**8.45**). In some flowering plants (begonias, orchids) the seeds are minute and the mature embryo shows little morphological differentiation.

Endosperm
In the majority of angiosperms two haploid polar nuclei occur in the embryo sac and their fusion with one of the sperm nuclei leads to the development of the triploid primary endosperm cell (**1.32**). However, in some species multiple polar nuclei occur so that the resulting endosperm is polyploid. In a few species fertilisation of the polar nuclei does not occur and in others division of the primary endosperm nucleus ceases very early. In many albuminous seeds (**8.43**) the nucellus and the inner integument degenerate as the endosperm develops in the maturing seed.

In most flowering plants the initial divisions of the primary endosperm nucleus are not followed by cytokinesis and a coenocytic nuclear endosperm develops (**2.5, 8.37**). However, subsequent free-wall formation leads to the cellularisation of the endosperm (**1.33, 8.32**). These walls resemble those in some *in vitro* cultured tissues (**2.62**) and in contrast to normal cell plate development neither microtubules nor Golgi bodies are apparently involved in their growth. The random wall formation results in some endosperm cells being multinucleate. In the mature fruit of *Cocos* (coconut, **8.46, 8.47**) the watery milk of the enclosed seed represents the remnants of the coenocytic cytoplasm while the white flesh results from its cellularisation.

Seed coat

The integuments of the ovule (8.33) develop into the seed coat (testa) after fertilisation (8.41, 8.48). The testa usually contains a hardened protective layer which may develop in the outer integument (8.49, 8.50) or from the inner integument. In the legumes the inner integument eventually degenerates but the outer epidermis of the remaining integument differentiates into a palisade layer of sclereids (8.50). At the hilum two layers of sclereids occur, with the outer forming from funicular tissue (8.49). A group of tracheids, which is probably concerned with water uptake during germination, also develops in this region (8.49). In cereals and grasses the grain (8.45, 8.51) is not a seed but rather a fruit (caryopsis), since the thin ovary wall is fused to the integument of the single ovule.

The fruit

Introduction

Following fertilisation the ovary develops into the fruit which encloses and protects the seeds (8.10, 8.27, 8.46, 8.48, 8.52). But some plants form parthenocarpic fruits without fertilisation of the ovules, as in *Musa* and *Ananas* (6.1, 8.53). A great diversity of fruits occurs but their typology will not be considered in detail here. Most fruits are termed simple since they are derived from a single ovary, which may be apocarpous (8.52) or syncarpous (1.6, 8.27, 8.54). Aggregate fruits are formed from several carpels of an individual flower (*Fragaria*, 8.55) while multiple fruits are formed from a number of flowers as in *Ananas* (8.53).

In addition to tissues directly derived from the ovary wall, adjacent accessory tissue may also contribute to the fruit body. In *Fragaria* (8.55) the swollen receptacle forms the centre of the fruit, whilst the inflorescence scales and swollen floral axis participate in the fruit of *Ananas* (8.53). The closely-crowded flowers of *Banksia* (8.56) develop into dehiscent woody fruits embedded in the swollen inflorescence axis (8.57) in which some secondary thickening occurs. In fruits developed from an inferior ovary (8.27), an outer layer of receptacle or perianth tissue is joined to the ovary wall. In *Malus* (apple) the flesh represents the swollen floral tube while the core is formed from the ovary (8.58).

The changes which occur in the fruit during maturation vary considerably in different plants. In grasses the ovary contains only a single seed (8.51) which at maturity is relatively small and the maturation changes in the ovary wall are mainly effected by vacuolation growth and cell sclerification. But in large-seeded or fleshy fruits (8.53, 8.58) active cell division accompanies

growth and in *Persea* (8.48) division occurs throughout its growth. In large and heavy fruits the original delicate flower stalk may become woody (8.58, 8.59) as further vascular tissue differentiates within it to meet the increased requirements of water and food stuffs for the developing fruit.

Fruit wall

The pericarp of fleshy fruits is often three-layered, with the exocarp comprising the skin, the mesocarp is represented by the flesh and the endocarp comprises the stone cells surrounding the seed (8.59). However, accessory tissues often contribute to the flesh as in the apple (8.58).

The cuticle-covered outer epidermis of the ovary wall is usually persistent in the fruit (8.60, 8.61) but the stomata are usually non-functional. Cork sometimes develops in some fruits and the patchy appearance of certain apple varieties is due to localised cork formation. The inner epidermis of a fruit adjacent to the loculus (8.60) is rarely cutinised. The sclerenchymatous endocarp of some fruits (8.59) originates from the inner epidermis and the adjacent ground tissue of the ovary wall. In many fleshy fruits the inner epidermis becomes secretory and in citrus fruits the flesh is formed by proliferating juice sacs which extend into the carpel locules (8.61).

The ground tissue of the ovary wall develops variously. In fruit walls which are dry at maturity much sclerenchyma develops (8.62, 8.63). In soft fruits (8.48, 8.53 to 8.55, 8.58) the ground tissue undergoes extensive vacuolation and predominantly consists of parenchyma penetrated by vascular strands linked to the pedicel. In the developing fruit the parenchyma often show maturation changes; the softening flesh of *Persea* (8.48) results from breakdown of the middle lamellae and the degeneration of the walls of this tissue.

Seed release from dry fruits

In plants where the fruit is not dispersed by animals (8.57, 8.62, 8.63) its principal function is seed protection. Indehiscent fruits, which are dry at maturity, often have thin walls which can be easily penetrated by the germinating embryo (8.64). Dehiscent fruits frequently have thick walls and break open to release the seeds (8.57, 8.62, 8.63, 8.65, 8.66). In legumes the pod (8.52) dehisces longitudinally along the suture of the carpel margins and also along the median vein (8.67). In many leguminous species a single layer of exocarpic fibres lie with their long axes more-or-less at right angles to that of the pod axis, while the endocarpic fibres run parallel to its axis.

As the pericarp dries, the exo- and endocarp contract in cross directions and the pod eventually ruptures (**8.67**). In some fruits the dehiscence zone is not related to carpellary sutures or vascular bundles and the fruit may break open by pores (**8.63**).

Forage fruits and seed dispersal

In dehiscent fruits the release and dispersal of its seed occurs while the fruit is still attached to the plant. Fleshy fruits are generally attractive as food for animals and the enclosed seeds are often protected from the animal's chewing and digestive juices by a sclerenchymatous endocarp (**8.59**), or thick-walled or mucilaginous seed coats (**1.6**). The seeds are later distributed in the animal's droppings. Small fleshy fruits (**8.55**), which are often non-aromatic, are mainly eaten by birds which have good colour vision but poor sense of smell. But larger aromatic fruits (**8.58**) are usually eaten by mammals, who have a well-developed sense of smell. Fleshy fruits also often fall to the ground where they are either eaten or rot and release the seeds.

Passive dispersal of fruits and seeds

Many fruits develop barbs or have sticky surfaces, and so become attached to passing animals and passively distributed. Dispersal of fruits and seeds by the wind is also common (**8.64**, **8.66**) but these rarely travel great distances. It has been calculated that even in the small and light fruits of *Taraxacum* (**8.10**) only about one per cent of the fruits are dispersed further than 10 kilometres. Some species have explosive fruits (**8.67**) with the ejected seeds being thrown up to several metres. Water dispersal is much less common but the large fruits of *Cocus* (coconut, **8.46**, **8.47**) and *Lodicea* are dispersed in the ocean currents with their enclosed seeds protected by a thick and sclerified endocarp. The fruit of the mangrove (*Rhizophora*) has a single seed which germinates *in situ* while the fruit is still attached to the tree. The radicle penetrates the fruit and grows up to 50cm over several months (**8.68**). The seedlings eventually fall off into the water or mud and develop into further mangrove forest (**7.6**, **8.69**).

8.1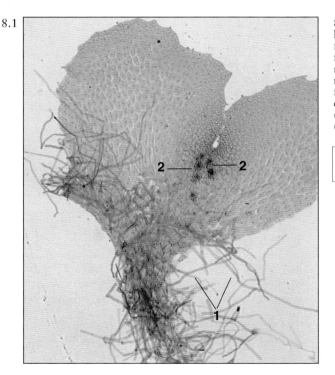

8.1 Abaxial surface view of a fern prothallus. When the haploid spores are released from the sporangia on the fern frond (cf., **1.7**) they germinate to form small dorsiventrally flattened gametophytes adhering to the soil by numerous rhizoids (1). The female reproductive organs (2) develop near the notch in the gametophyte and the eggs are fertilised by flagellate sperms; the zygote is diploid and the embryo rapidly develops into a young fern plant (sporophyte) independent of the short-lived gametophyte. (LM x 30.)

1 Rhizoids	2 Female reproductive organs

8.2 LS of the cycad ovule of *Zamia* showing a pair of large archegonia. In seed plants the megasporangium (nucellus, 1) is invested by sterile integuments and the whole structure comprises the ovule (cf., **1.5**). The female gametophyte (2) develops *in situ* from a single haploid megaspore (cf., **1.32**) and, at its micropylar end, several vestigial female sex organs develop. Essentially each consists of a single massive egg (3) embedded in female gametophytic tissue. Fertilisation is effected by multiflagellate sperm. (LM x 25.)

1 Nucellus	3 Egg
2 Female gametophyte	

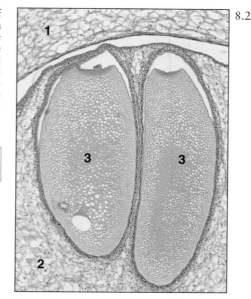

8.2

8.3

8.3 Unisexual flowers of the dicotyledon *Begonia sempervirens*. Both sexes occur in the same plant: the irregular (zygomorphic) male flower shows two large and two small perianth members and a central cluster of yellow stamens (1). In the regular actinomorphic female flower five or more perianth members surround an inferior ovary terminated by convoluted stigmas (2).

1 Stamens	2 Stigmas

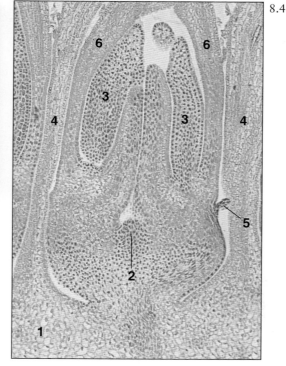

8.4

8.4 LS of an immature floret of the dicotyledon *Helianthus* (sunflower). The numerous florets are condensed on a capitulum (1) to make the large sunflower head. The floret has a central inferior ovary with a single basal ovule (2). Anthers (3), bract (4), sepal (5), petal (6). (LM x 70.)

1 Capitulum	4 Bract
2 Basal ovule	5 Sepal
3 Anther	6 Petal

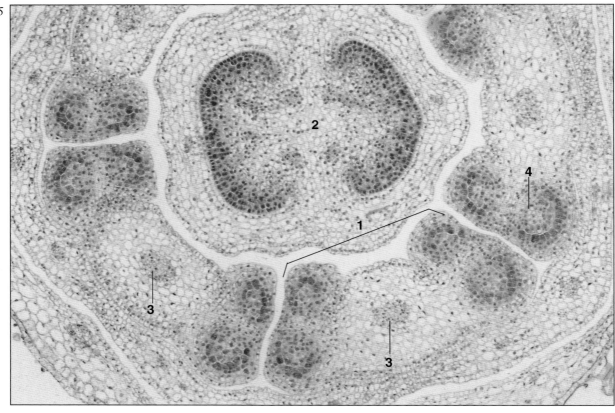

8.5 TS of the actinomorphic flower bud of *Solanum tuberosum* (potato). This dicotyledon has five anthers (1) and a bilocular, superior ovary (2). Connective vascular bundle (3), pollen sac (4). (LM x 110.)

1 Anther	3 Vascular bundle
2 Superior ovary	4 Pollen sac

8.6 Dicotyledonous actinomorphic flowers of *Aquilegia* (columbine). Each is composed of five orange-red petaloid sepals (1) alternating with five yellow petals (2) that terminate in orange-red, nectar-secreting spurs (3). In the centre of each flower numerous stamens and five free carpels occur. Long-tongued bumble bees visit the flowers for nectar and pollen.

1 Petaloid sepals	3 Nectar-secreting spurs
2 Petals	

8.7 *Zygomorphic flowers of the dicotyledon *Cytisus scoparius* (broom). In the flower bud the tubular calyx (1) does not enclose the petals, but the large standard petal (2) surrounds the lateral wings (3) and abaxial pair of adherent keel petals (4). The stamens and single carpel enclosed within the keel would be exposed by large visiting bees.

1	Calyx	3	Lateral wings
2	Standard petal	4	Keel petals

1	Male catkins
2	Female catkins

8.8 Male and female inflorescences of the dicotyledon *Alnus glutinosa* (alder). The male catkin (1) bears numerous groups of three florets borne on short branches from the inflorescence axis. Each floret is reduced to two vestigial perianth members and four stamens with freely exposed anthers. The as yet unopened female catkins (2) also bear groups of three florets, each has a bicarpellary ovary with two long, curved styles pollinated by wind-borne grains.

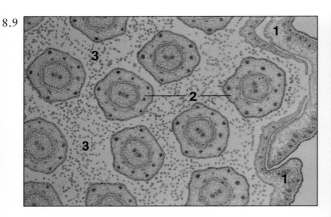

8.9 TS of an immature capitulum of the dicotyledon *Taraxacum officinale* (dandelion). Numerous florets (cf., **8.4**) are surrounded by large leafy bracts (1) borne on the margin of the capitulum. Each floret has a tubular corolla (2), while the numerous external hairs (3) represent the calyx (LM x 25.)

1	Leafy bracts	3	Calyx (pappus)
2	Corolla		

8.10 Capitulum of the dicotyledon *Taraxacum officinale* (dandelion) bearing numerous fruits. These are wind-dispersed and each shows a terminal parachute-like pappus formed from the calyx (cf., **8.9**). Despite pollination by the many nectar-collecting insects, fertilisation is rare. Instead a diploid egg develops apomictically into the embryo of the single-seeded fruit.

8.11 Large regular flower of the dicotyledon *Papaver* (poppy). The four crumpled petals (1) show bluish-black pollen shed onto their adaxial surfaces. The copious pollen from the massed anthers (2) attracts insects which settle on the wide stigmatic disc (3) and thus effect cross pollination.

1	Petals	3	Stigmatic disc
2	Anthers		

8.12A

8.12 Flowers of the dicotyledon *Ranunculus* (buttercup) under normal (**A**) and UV illumination (**B**). Note the regular arrangement of the five petals (1), the abundant stamens (2) and the numerous free, superior, carpels (3). In **B**, the nectar guides on the adaxial surfaces of the petals are revealed; these are visible to bees and hover flies and apparently guide them to the nectaries at the base of the petals. Pollen is shed onto the petals and is easily picked up by insects. (*Copyright of T. Norman Tait.*)

1	Petals
2	Stamens
3	Carpels

8.12B

8.13 TS of the immature capitulum of the dicotyledon *Helianthus* (sunflower). Each floret (cf., 8.4) is subtended by a bract (1) and shows a tubular corolla, five anthers (2), and style terminated by a bifid stigma (3). The sepals and inferior ovary lie beneath the level of this section. (LM x 75.)

1	Bract
2	Anthers
3	Bifid stigma

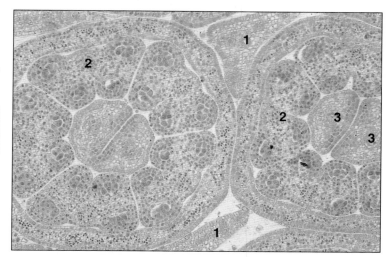

8.14

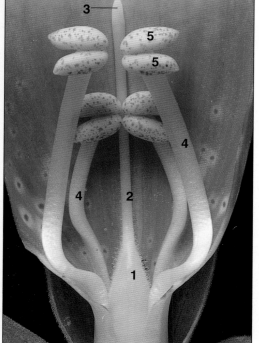

8.15

8.14 Close-up of the reproductive organs of the flower of the monocotyledon *Crocus*. The anthers (1) have dehisced longitudinally and reveal the massed pollen grains; whilst the long style (2) branches into three expanded stigmas (3). Long-tongued bees and butterflies visit the flower for the nectar secreted at the base of the perianth and brush against the copious pollen.

1	Anther	3	Stigmas
2	Style		

8.15 Close-up of the flower of the dicotyledon *Digitalis purpurea* (foxglove). The irregular tubular corolla (cf., 5.7) has been split open to reveal the superior ovary (1) with a long style (2) and terminal stigma (3). There are four stamens with their filaments (4, two short and two long) joined to the corolla tube. Each bears a two-lobed anther (5). Pollination is by bumble-bees visiting for the nectar secreted at the base of the ovary.

1	Ovary	4	Filaments
2	Style	5	Anthers
3	Stigma		

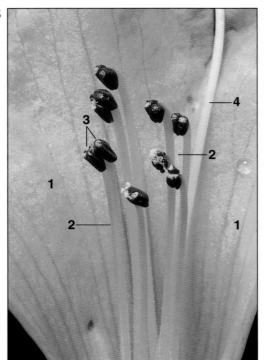

8.16

8.16 Close-up of the flower of the dicotyledon *Rhodo-dendron*. The corolla tube (1), has been split open to reveal the ten stamens whose filaments (2) are of varying lengths. The two-lobed anthers exude their pollen from terminal pores (3); the pollen tetrads adhere to each other, forming white masses that stick to bumble-bees visiting for nectar secreted at the base of the corolla. Style (4).

1	Corolla tube	3	Terminal pores
2	Filaments	4	Style

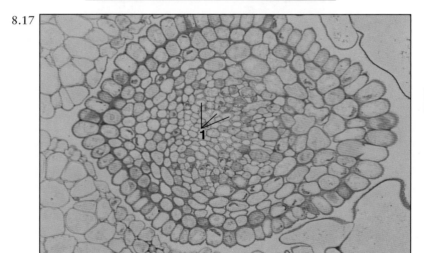

8.17

8.17 TS of the filament of an immature anther of the dicotyledon *Sinapis*. Note its central vascular strand which transports nutrients to the developing pollen. Tracheary elements (1). (G-Os, LM x 290.)

1	Tracheary elements

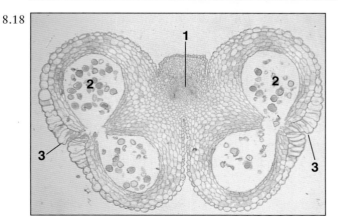

8.18

8.18 TS of the dehisced anther of the monocotyledon *Lilium* (lily). The two anther lobes are joined by the connective tissue containing a vascular bundle (1). The wall between the two pollen sacs has broken down and the pollen grains (2) released by the dehiscence of the anther walls at the stomia (3). The hypodermal walls (except at the stomia) are elaborately thickened (cf., **8.20**). (Polarised LM x 25.)

1	Vascular bundle	3	Stomium
2	Pollen grains		

8.19 TS of immature pollen sacs in a flower bud of the dicotyledon *Solanum tuberosum* (potato). Each anther lobe (cf., **8.5**) contains two pollen sacs in which the hypodermis has undergone periclinal divisions forming vacuolated parietal tissue (1) and several layers of densely-staining tapetal cells (2). From the core of sporogenous cells (3) the haploid pollen grains are derived after meiosis. (Phase LM x 205.)

1	Parietal tissue
2	Tapetal cells
3	Sporogenous cells

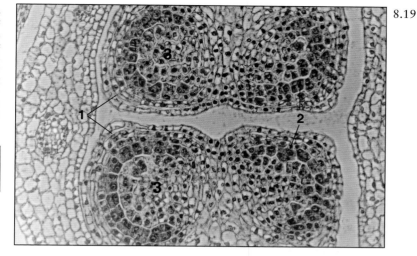

8.20 TS showing details of the anther wall of the monocotyledon *Lilium* (lily). This is a mature anther (cf., **8.18**) and shows the large but unthickened epidermal cells of the stomium (1). These contrast with the hypodermal cells (2) whose walls show anticlinal bands of cellulosic thickening. Pollen grains (3). (Polarised LM x 75.)

1	Stomium
2	Hypodermal cells
3	Pollen grains

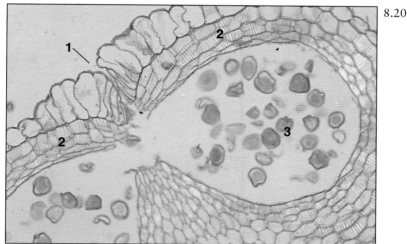

8.21 TS showing detail of pollen mother cells in monocotyledon *Lilium* (lily) anther. Their large nuclei are in early prophase of the first meiotic division and the chromosomes appear as thread-like structures separated from the cytoplasm (1) by the nuclear envelope (2). Tapetal nuclei (3). (Phase LM x 500.)

1	Cytoplasm
2	Nuclear envelope
3	Tapetal nuclei

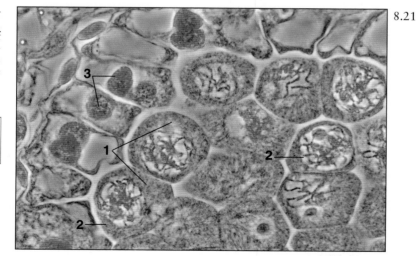

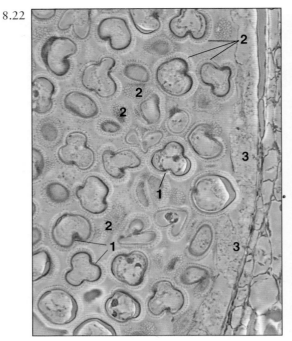

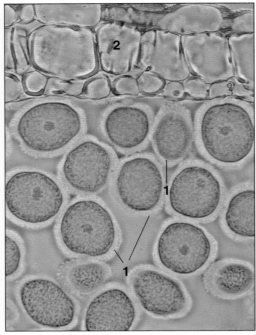

8.22 Sectioned anther of the dicotyledon *Sinapis* showing pollen grains undergoing cytokinesis after meiosis. Wall formation in the pollen mother cell is of the simultaneous pattern with the wall furrows (1) developing centripetally to form four pollen grains. Note the patterned exine (2) and the degenerate remains of the tapetum (3). (G-Os, Phase LM x 460.)

1 Wall furrows	3 Tapetum
2 Patterned exine	

8.23 Anther of the dicotyledon *Sinapis* showing individual pollen grains. These have almost separated from the tetrads (cf., **8.22**) and are considerably enlarged; each grain is surrounded by a thick, patterned exine (1). Fibrous anther wall (2). (G-Os, Phase LM x 460.)

1 Patterned exine	2 Anther wall

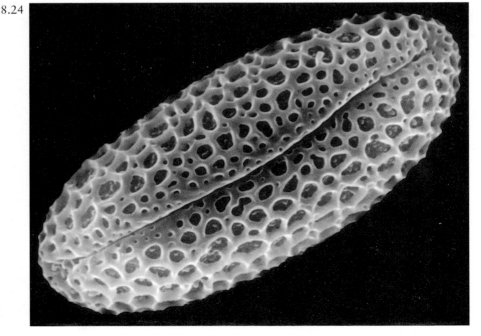

8.24 Mature pollen grain of the dicotyledon *Salix* (willow). This grain has a reticulate exine and one of its three longitudinal grooves (colpi) is visible. The grains are insect-dispersed and at germination the pollen tube emerges from one of the colpi. (SEM x 6,400.) (*Copyright of Dr James H. Dickson.*)

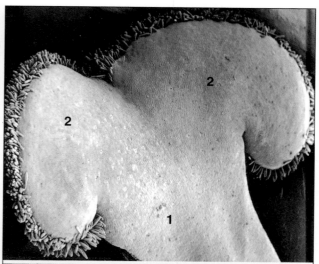

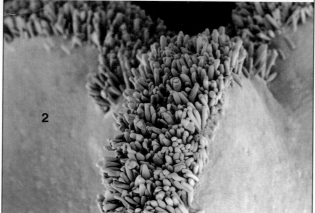

8.25 Mature pollen grain of the dicotyledon *Malva* (mallow). Note that the exine is coarsely ornamented with large spikes. The grains are insect-dispersed and at germination the pollen tube emerges through one of the numerous germ pores. (SEM x 2,700.) (*Copyright of Dr James H. Dickson.*)

1 Style	2 Stigmatic ridges

8.26 Surface detail of the stigma of the monocotyledon *Tulipa* (tulip). In **A**, the style (1) expands into three ridges (2) reflecting the trilocular nature of its ovary. The surfaces of these ridges (**B**) bear closely crowded, short glandular trichomes in which pollen grains become enmeshed. (SEM, **A** x 18, **B** x 36.)

8.27 TS and LS of the fruit of the monocotyledon *Narcissus* (daffodil). **A** shows its trilocular nature and the axial placentation of the seeds (1). **B** illustrates the inferior fruit lying at the base of the withered corolla tube (2). Bract (3).

1 Seeds
2 Corolla tube
3 Bract

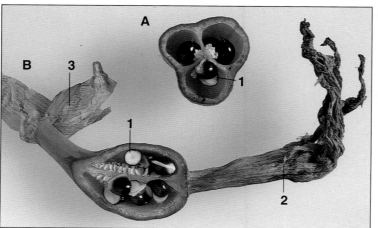

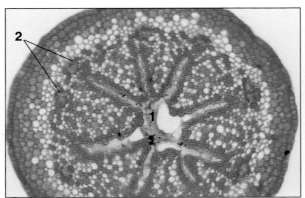

8.28 TS of the style of the dicotyledon *Rhododendron*. The central channel and radiating arms contain mucilage (1) secreted by the lining epithelial cells; the growing pollen tubes are nourished by this secretion (cf., **8.29**). The stylar wall contains a number of longitudinal vascular bundles (2). (LM x 65.)

1 Mucilage	2 Vascular bundles

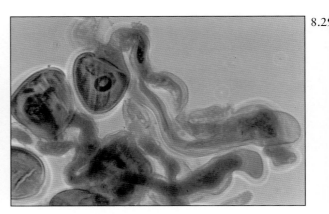

8.29 Details of germinating pollen of the monocotyledon *Narcissus* (daffodil). These grains have germinated *in vitro* in agar with 7% sucrose and show well-developed, convoluted pollen tubes; several nuclei are also evident (Phase LM x 370.)

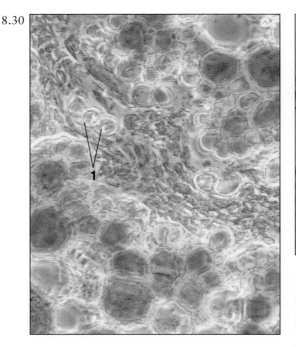

8.30 TS of the style of the dicotyledon *Rhododendron* showing growing pollen tubes. Note within its mucilaginous interior (cf., **8.28**) several transversely-sectioned pollen tubes (1). (Phase LM x 460.)

1 Pollen tubes

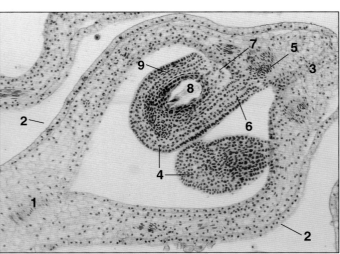

8.31 TS of a carpel of the dicotyledon *Caltha palustris* (marsh marigold). This shows its leaf-like nature with the dorsal suture (1) corresponding to a midrib whilst the blade is folded with its abaxial surface (2) outermost. The two leaf margins are fused at the ventral suture (3) and the inverted ovules (4) are joined to the placenta (5) by a short funiculus (6). Micropyle (7), embryo sac (8), integuments (9). (LM x 75.)

1 Dorsal suture	6 Funiculus
2 Abaxial surface	7 Micropyle
3 Ventral suture	8 Embryo sac
4 Ovule	9 Integument
5 Placenta	

160

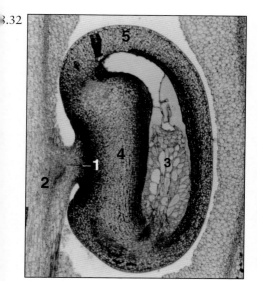

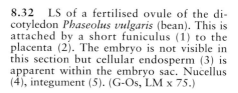

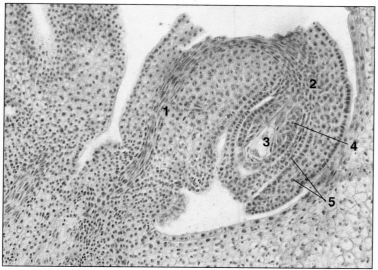

8.32   LS of a fertilised ovule of the dicotyledon *Phaseolus vulgaris* (bean). This is attached by a short funiculus (1) to the placenta (2). The embryo is not visible in this section but cellular endosperm (3) is apparent within the embryo sac. Nucellus (4), integument (5). (G-Os, LM x 75.)

1	Funiculus
2	Placenta
3	Cellular endosperm
4	Nucellus
5	Integument

8.33   LS of anatropous ovules of the monocotyledon *Iris*. The medianly-sectioned ovule arises from the central axis of a trilocular ovary (cf., **8.27**) with the funiculus, containing a vascular strand (1), running from the placenta to the chalaza (2) at the base of the embryo sac (3). Nucellus (4), integuments (5). (LM x 125.)

1	Vascular strand	4	Nucellus
2	Chalaza	5	Integuments
3	Embryo sac		

8.34   LS of an ovule primordium in the monocotyledon *Lilium* (lily). At the micropolar pole of the nucellus (1) a single hypodermal cell has enlarged and its nucleus (2) is about to undergo meiosis. Integument (3), placenta (4). (LM x 330.)

1	Nucellus
2	Hypodermal nucleus
3	Integument
4	Placenta

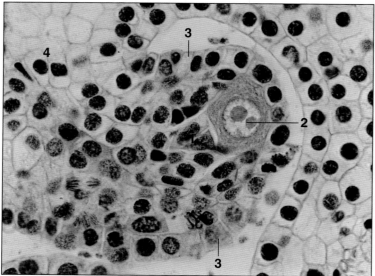

8.35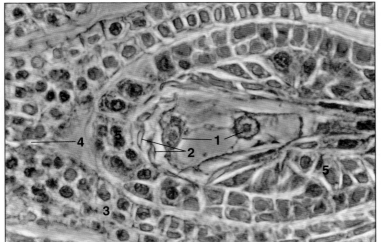

8.35 LS of a semi-mature ovule of the monocotyledon *Iris*. The embryo sac contains two functional nuclei (1) but several degenerate nuclei (2) represent the non-functional megaspores and nucellar cells crushed by the expanding embryo sac. Integument (3), micropyle (4), nucellus (5). (LM x 530.)

1	Functional nuclei
2	Degenerate nuclei
3	Integument
4	Micropyle
5	Nucellus

8.36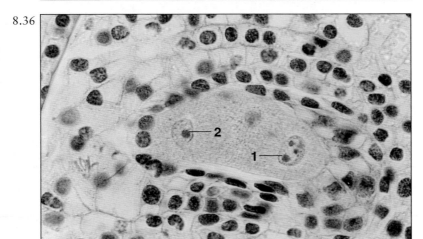

8.36 LS of an immature embryo sac of the monocotyledon *Lilium* (lily). The megaspore mother cell (cf., **8.34**) has divided meiotically to give four nuclei, but without wall formation. Three of these nuclei have fused to give a triploid nucleus (1) at the chalazal end of the embryo sac while a haploid nucleus (2) remains at the other pole. Subsequently the egg, two synergids and one polar nucleus form from the latter, while the triploid nucleus gives rise to three antipodal and one polar nucleus. (LM x 330.)

1	Triploid nucleus
2	Haploid nucleus

8.37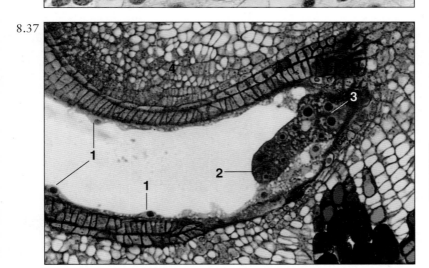

8.37 LS of a very young embryo of the dicotyledon *Phaseolus vulgaris* (bean). This is located at the micropylar pole of the large embryo sac (cf., **8.32**) and nuclei (1) of the coenocytic endosperm line the walls. The embryo is terminated by the proembryo (2) whilst the cylindrical suspensor (3) is distended at its micropylar end. Nucellus (4). (G-Os, LM x 330.)

1	Endosperm nuclei
2	Proembryo
3	Suspensor
4	Nucellus

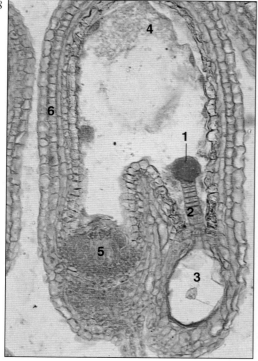

8.38 Immature seeds of the dicotyledon *Capsella bursa-pastoris* (shepherd's purse). The longitudinally-sectioned seed shows a globular proembryo (1) attached to a filamentous suspensor (2) with its swollen basal cell (3) terminating at the micropylar end of the embryo sac (cf., **1.33**). Coenocytic endosperm (4), hypertrophied nutritive nucellar tissue (5), testa (6). (LM x 125.)

1	Proembryo	4	Coenocytic endosperm
2	filamentous suspensor	5	Nucellar tissue
3	Basal cell	6	Testa

8.39 Detail of the micropylar pole of the embryo in the dicotyledon *Phaseolus vulgaris* (bean). The nuclei of the basal cells of the suspensor (1) undergo numerous rounds of DNA replication, which is not accompanied by mitosis, and very large highly polytene nuclei (2) develop. Nucellus (3). (G-Os, LM x 330.

1	Suspensor	3	Nucellus
2	Polytene nuclei		

8.40 Immature seed of the dicotyledon *Phaseolus vulgaris* (bean). The section cuts the embryo transversely through the hypocotyl: note the differentiation of pith (1), vascular tissue (2), and cortex (3). The peripheral endosperm layer (4) has become detached from the wall of the embryo sac. (G-Os, LM x 85.)

1	Pith
2	Vascular tissue
3	Cortex
4	Endosperm layer

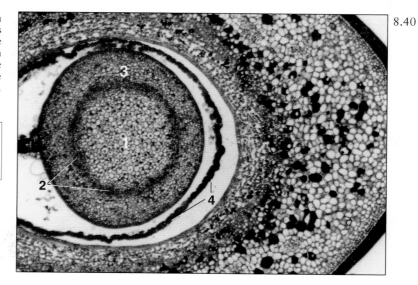

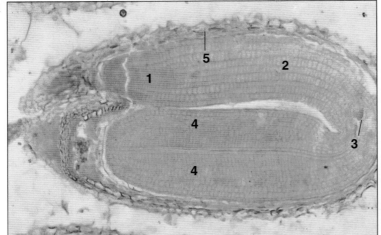

8.41 LS of the mature seed of the dicotyledon *Capsella bursa-pastoris* (shepherd's purse). The embryo completely fills the embryo sac and all the endosperm has been absorbed (cf., **1.33**). A wide cylindrical radicle (1, with its apex at the micropyle) is continuous with the hypocotyl (2). This is terminated by a small plumular apex (3) on either side of which arise a pair of swollen cotyledons (4) lying parallel to the radicle-hypocotyl axis. The curved nature of its embryo sac (cf., **8.38**) results in the bent shape of the embryo. Testa (5).(LM x 100.)

1	Radicle
2	Hypocotyl
3	Plumular apex
4	Cotyledons
5	Testa

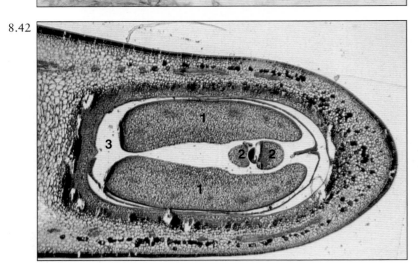

8.42 Semi-mature seed of the dicotyledon *Phaseolus vulgaris* (bean). The embryo is cut obliquely and shows the paired cotyledons (1) whilst the primordia of the first pair of foliage leaves (2) have already been formed by the shoot apex (cf. **4.1**). Embryo sac (3). (G-Os, LM x 30.)

1	Cotyledons
2	Foliage leaves
3	Embryo sac

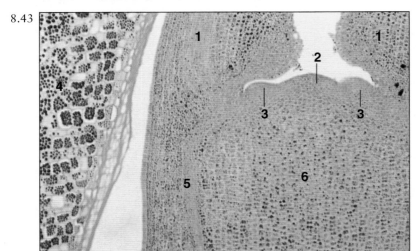

8.43 LS of the mature plumule in the seed of the dicotyledon *Ricinus communis* (castor oil). Note the paired cotyledons (1) on either side of the shoot apex (2) and the first foliage leaf primordia (3). The seed contains copious supplies of endosperm (4) to nourish the germinating embryo. Procambium (5), pith (6). (LM x 70.)

1	Cotyledon
2	Shoot apex
3	Foliage leaf primordia
4	Endosperm
5	Procambium
6	Pith

8.44 LS of the mature radicle in the seed of the dicotyledon *Ricinus communis* (castor oil). Note the root cap (1), apical meristem (2), procambium (3) and endosperm (4). (LM x 80.)

1	Root cap
2	Apical meristem
3	Procambium
4	Endosperm

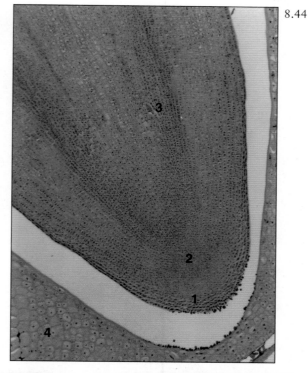

8.44

8.45

8.46

8.46 Fruiting crown of the monocotyledon *Cocos nucifera* (coconut palm). These large monocotyledons frequently grow along tropical beaches in Asia and Polynesia; they bear clusters of large one-seeded fruits that fall to the ground when ripe and are often dispersed great distances by the sea currents (cf., **8.47**).

8.45 Germinating grain of the monocotyledon *Zea mays* (maize). Note the long radicle with its dense covering of root hairs (1) and the remains of the coleorhiza (2) at its base. The latter covered the root prior to germination, while the coleoptile (3) invests and protects the plumule. Seminal root (4). (*Copyright of T. Norman Tait.*)

1	Root hairs	3	Coleopeptile
2	Coleorhiza	4	Seminal root

165

Fruit of *Cocos nucifera* (coconut palm). This was washed up onto an island many kilometres from the nearest coconut palms. The original smooth exocarp (cf., **8.46**) of the fruit has rotted to reveal the thick fibrous mesocarp enclosing the hard endocarp (coconut shell). The single seed within the fruit is enclosed by a thin brown testa (not distinguishable in this illustration) which is lined by a layer of cellular endosperm. This now grey decayed tissue normally forms the edible white flesh. The small embryo is embedded in this flesh but is not evident in the illustration.

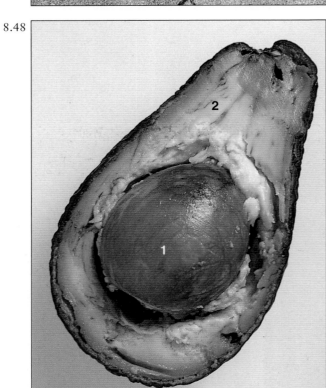

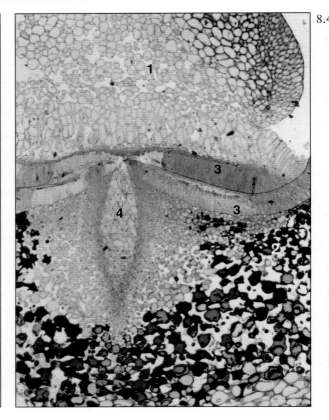

8.48 LS of the fruit of the dicotyledon *Persea americana* (avocado pear). This forms from a superior unilocular ovary bearing a single pendulous ovule that develops into the large central seed (1). The edible parenchymatous fruit wall (2) contains up to 30% of oils that initially accumulate in oil sacs. In the mature fruit the tissue degenerates and the flesh is buttery.

1 Seed	2 Parenchymatous fruit wall

8.49 TS through the hilum of the semi-mature seed of the dicotyledon *Phaseolus vulgaris* (bean). The funiculus (1) is confluent with the testa (2); at their junction two layers of columnar epidermal cells are apparent (3) that subsequently form macrosclerides (cf., **8.50**). A compact group of tracheids later develops in the centre of the hilum from a lens-shaped group of cells (4). (G-Os, LM x 80.)

1 Funiculus	3 Epidermal cells
2 Testa	4 Potential tracheids

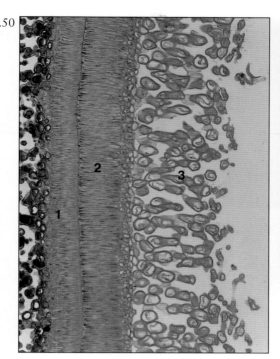

8.51 Harvested cob of *the* monocotyledon *Zea mays* (maize). This matures several months after fertilisation from an inflorescence axis bearing numerous female flowers. The cob is invested by bracts (peeled back) that cover the closely-crowded, swollen grains (fruits). Note the withered tassel representing the collective remains of the pendulous, thread-like style borne by each flower; the styles may be up to 25 cm long and remain receptive to pollination for up to two weeks.

1	Inner macrosclereids
2	Outer macrosclereids
3	Spongy tissue

8.50 LS through the hilum of the mature seed of the dicotyledon *Phaseolus vulgaris* (bean). Note the two layers of macrosclereids (cf., **8.49**); the inner (1) is derived from the epidermis of the integument whilst the outer layer (2) forms from the funicular epidermis; the spongy tissue (3) is hypodermal in origin. (LM x 110.)

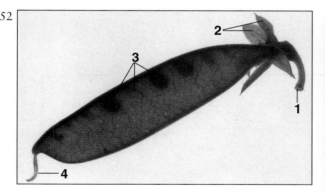

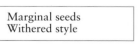

8.52 Young pod of the dicotyledon *Pisum sativum* (pea). In legumes the fruit develops from a superior apocarpous ovary. Note the pedicel (1), green sepals (2), swollen fruit with a row of marginal seeds faintly visible (3) and withered remains of the style (4).

| 1 | Pedicel | 3 | Marginal seeds |
| 2 | Green sepals | 4 | Withered style |

8.53 The multiple fruit of the monocotyledon *Ananas comosus* (pineapple). In internal view the fleshy inflorescence axis (1) is shown cut lengthwise and numerous swollen, coalesced fruitlets arise laterally from it. Each develops without fertilisation from an inferior ovary. In external view the numerous, spirally arranged fruitlets (2) are visible. After flower production the inflorescence apex reverts to vegetative growth and the leafy crown (3) is used to propagate this crop vegetatively.

1	Inflorescence axis
2	Fruitlets
3	Leafy crown

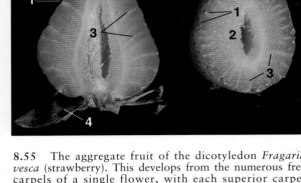

8.54 TS of the fruit of the dicotyledon *Cucurbita* (cucumber). This develops from an inferior syncarpous ovary of three fused carpels with parietal placentae (1). The numerous seeds (2) are enclosed in a pulpy parenchymatous endocarp; this is surrounded by the firmer flesh of the mesocarp (3) in which a ring of bicollateral vascular bundles (cf., **6.11**) occurs. The epicarp forms the green skin of the fruit.

8.55 The aggregate fruit of the dicotyledon *Fragaria vesca* (strawberry). This develops from the numerous free carpels of a single flower, with each superior carpel containing a single ovule. The resultant indehiscent fruitlets (pips, 1) are embedded in the hypertrophied floral receptacle (2) to form a succulent aggregate fruit. Note the white longitudinal vascular strands (3) which branch to supply the individual fruitlets. Calyx (4).

1	Parietal placentae	3	Mesocarp
2	Seeds		

1	Fruitlets (pips)	3	Vascular strands
2	Floral receptacle	4	Calyx

8.56 Inflorescence of the dicotyledon *Banksia* bearing closely-crowded flowers. The flowers of this indigenous Australasian genus are bird-pollinated and subsequently the ovaries become embedded in the woody inflorescence axis to form a cone-like structure (cf., **8.57**).

8.57 Cone-like fruitification of the dicotyledon *Banksia*. This woody genus forms closely-crowded sclerified fruits embedded in the swollen inflorescence axis that undergoes some secondary thickening. The fruits dehisce, usually following a bush fire, along a horizontal suture to release the seeds.

8.58 Fleshy fruit of the dicotyledon *Malus sylvestris* (apple). The core of the apple (with its tough sclereids) forms from an inferior syncarpous ovary with axile placentation. The parenchymatous flesh represents the greatly enlarged floral tube which surrounded the ovary. In cross-section of the fruit, four of the five ovary compartments contain seeds (1). In the inner flesh there is a ring of vascular bundles (2) that supplied the five sepals and five petals. In longitudinal view the pedicel (3) and the withered remains (4) of calyx, stamens and styles are visible.

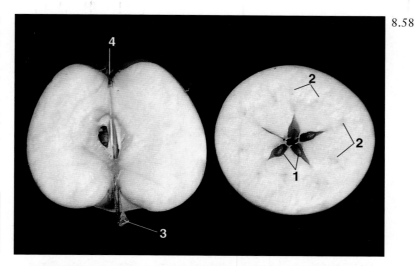

1	Seeds
2	Vascular bundles
3	Pedicel
4	Remains of calyx, stamens, styles

8.59 LS of the fruit of the dicotyledon *Prunus* (peach). This develops from a unilocular superior ovary. The stony inner endocarp (1) encloses a single seed. The extensive mesocarp (2) is fleshy and succulent and is enclosed by the thin skin (epicarp, 3). Pedicel (4).

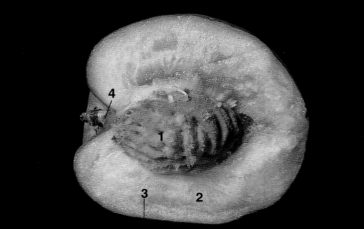

1	Stony endocarp
2	Fleshy mesocarp
3	Epicarp (skin)
4	Pedicel

8.60 LS of a young fruit of the dicotyledon *Phaseolus vulgaris* (bean). The outer epidermis of the pod bears numerous trichomes and a cuticle is present, but the inner surface (1) is hairless and lacks a cuticle. Vascular strand (2), immature seed (3), cellular endosperm (4). (G-Os, LM x 80.)

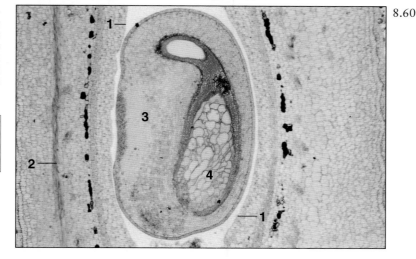

1	Inner surface of pod
2	Vascular strand
3	Immature seed
4	Cellular endosperm

8.61

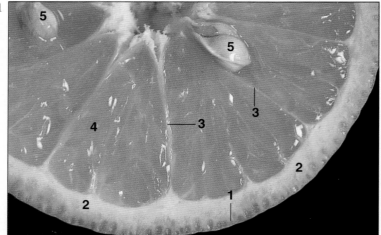

8.61 TS of the fruit of the dicotyledon *Citrus sinensis* (orange). This develops from a superior ovary of ten united carpels with axile placentation. The peel of the fruit consists of the orange leathery exocarp (1, containing numerous oil glands) and white aerenchymatous mesocarp (2). The endocarp forms a thin layer internal to the mesocarp and also the radial partitions (3) between the locules. Ingrowths from the endocarp develop into the juice sacs (4) which pack the interiors of the locules. Seeds (5).

1	Exocarp with oil glands
2	Mesocarp
3	Radial partitions (endocarp)
4	Juice sacs
5	Seeds

8.62

8.62 Pendulous fruits of the dicotyledon *Eucalyptus calophylla*. This indigenous Australasian genus has woody capsules that dehisce at their tips to release the seeds. The woody fruit develops from an inferior, syncarpous ovary to which the surrounding floral receptacle is united.

8.63

8.63 Ripening fruits of the dicotyledon *Papaver* (poppy). The woody capsule forms from a superior syncarpous ovary. In the mature fruit the stigmatic lobes bend upwards from their previous position (cf., **8.11**) and uncover a ring of pores (not visible) in the capsule wall. As the capsule is blown by the wind, the small seeds gradually sift through the pores.

8.64   Immature fruit of the dicotyledon *Acer pseudoplatanus* (sycamore). This forms from a superior bicarpellary ovary whose walls develop two prominent wings, each enclosing a single seed. At maturity the fruit abscises and its wings cause the fruit to spin downwards to the ground where it separates into two non-dehiscent segments, each containing a single seed. Pedicel (1), withered remnants of style (2), position of seeds within pericarp (3).

1	Pedicel
2	Style remnants
3	Position of seeds

8.64

8.65   Dehisced fruit of the monocotyledon *Iris*. This forms from an inferior trilocular ovary that at maturity is dry and dehisces into three valves to liberate its red seeds. (*Copyright of T. Norman Tait.*)

8.65

8.66   Dehiscing fruit of the dicotyledon *Epilobium hirsutum* (willow herb). The inferior ovary has four locules with axile placentae bearing numerous ovules; the mature fruit dehisces from the top downwards into four valves. The plumed seeds are dispersed in the wind. (*Copyright of T. Norman Tait.*)

8.66

8.67

8.68

8.67 Dehisced pods of the dicotyledon *Cytisus scoparius* (broom). The old pods have dehisced longitudinally along both sutures; as the dry pods split open they become twisted and violently eject the seeds.

8.68 Viviporous seedlings of the dicotyledon *Rhizophora* (mangrove). The embryo germinates within the fruit (1, note the persistent basal calyx) whilst the latter is still attached to the mangrove tree. The elongate seedling axis comprises a long radicle-hypocotyl axis (2) but the plumule still lies within the fruit. The seedlings reach 30–50 cm. before falling from the tree and rooting in the mud flats (cf., **8.69**). (*Copyright of T. Norman Tait.*)

1 Fruit	2 Radicle-hypocotyl axis

8.69

8.69 Newly-established colony of the dicotyledon *Rhizophora* (mangrove). The viviporous seedlings (cf., **8.68**) root in the estuarine mud or shallow water and form new colonies. Note the numerous seedlings in the foreground. (*Copyright T. Norman Tait.*)

172

Selected bibliography

Bell, A.D. (1991) *An Illustrated Guide to Flowering Plant Morphology.* Oxford University Press, Oxford, p.341. ISBN 0-19-854219-4.

Burgess, J. (1985) *An Introduction to Plant Cell Development.* Cambridge University Press, Cambridge, p.239. ISBN 0-521-31611-1.

Endress, P.K. (1994) *Diversity and Evolutionary Botany of Tropical Flowers.* Cambridge University Press, Cambridge, p.511. ISBN 0-521-42088-1.

Esau, K. (1977) *Anatomy of Seed Plants.* John Wiley & Sons, New York, p.550.

Fahn, A. (1990) *Plant Anatomy*, 4th edn. Pergamon Press, Oxford, p.588. ISBN 0-08-037490-5 (hardcover)/0-08-037491-3 (flexicover).

Foster, A.S. and Gifford, E.M. (1974) *Comparative Morphology of Vascular Plants*, 2nd edn. W.H. Freeman & Co, San Francisco, p.751. ISBN 0-7167-0712-8.

Heywood, V.H. (ed) (1993) *Flowering Plants of the World.* BT Batsford, London, p.335. ISBN 0-7134-7422-X (flexicover).

Hickey, M. and King, C.J. (1981) *100 Families of Flowering Plants.* Cambridge University Press, Cambridge, p.567. ISBN 0-521-2989-1.

Ingrouille, M. (1992) *Diversity and Evolution of Land Plants.* Chapman & Hall, London, p.340. ISBN 0-412-44230-2.

Lyndon, R.F. (1990) *Plant Development: The Cellular Basis.* Unwin Hyman, London, p.342. ISBN 0-04-581032-X (hardcover)/0-04-581033-8 (flexicover).

Mauseth, J.D. (1988) *Plant Anatomy.* Benjamin/Cummings Publishing Company, Menlo Park, California, p.560. ISBN 0-8053-4570-1.

Mauseth, J.D. (1995) *Botany – An Introduction to Plant Biology*, 2nd edn. Saunders College Press, Philadelphia. ISBN 0-030-96842-9 (hardcover).

Raghavan, V. (1986) *Embryogenesis in Angiosperms.* Cambridge University Press, Cambridge, p.303. ISBN 0-521-26771-4.

Rudall, P. (1992) *Anatomy of Flowering Plants*, 2nd edn. Cambridge University Press, Cambridge, p.110. ISBN 0-521-42154-3.

Sporne, K.F. (1974) *The Morphology of Angiosperms.* Hutchinson & Co, London, p.207. ISBN 0-09-120610-3 (hardcover)/0-09-912611-1 (flexicover).

Sporne, K.F. (1974) *The Morphology of Gymnosperms*, 2nd edn. Hutchinson & Co, London, p.216. ISBN 0-09-077151-6 (hardcover)/0-09-077152-4 (flexicover).

Taiz, L. & Zeiger, E. (1991) *Plant Physiology.* Benjamin/Cummings Publishing Company, Menlo Park, California, p.559. ISBN 0-8053-0153-4.

Weberling, F. (1989) *Morphology of Flowers and Inflorescences.* Cambridge University Press, Cambridge, p.405. ISBN 0-521-25134-6 (hardcover)/0-521-43832-2 (flexicover).

Glossary

Abaxial surface The surface of leaf or bud primordium remote from the shoot apex or stem (cf., adaxial). The lower surface of a bifacial leaf.

Abscission zone The region at the base of a leaf, or other deciduous organ, which ruptures along the abscission layer to reveal a protective layer of cork covering the scar.

Abscission layer The layer of cells whose breakdown leads to the shedding of a deciduous organ.

Accessory bud A bud additional to the main axillary bud.

Accessory tissue The tissues of the flower or inflorescence which may be associated with the ovary in fruit development.

Acropetal differentiation Forming from the base of an organ towards its apex.

Actinomorphic A regular, radially symmetrical flower (cf., zygomorphic).

Adaxial surface The surface of a leaf or bud adjacent to the shoot apex or stem (cf., abaxial).

Adventitious Plant organs arising from unusual locations; as in buds developing on roots or leaves, roots forming on leaves and stems.

Aerenchyma Parenchymatous tissue with very large intercellular spaces.

Aerobic respiration Organisms utilising molecular oxygen for respiration.

Aggregate fruit A fruit formed from several or numerous carpels of a single flower, e.g. strawberry.

Albuminous seed A seed which at maturity contains endosperm to nourish the germinating embryo.

Alga A member of the large group of non-vascular plants which are thalloid and predominantly aquatic.

Alternation of generations In all land plants the diploid sporophyte generation is different in form to the haploid gametophyte (heteromorphic alternation). In vascular plants the sporophyte generation is dominant in contrast to the bryophytes where the gametophyte is dominant.

Amyloplast A non-coloured plastid storing large quantities of starch in its stroma.

Amphicribral vascular bundle A vascular bundle with a xylem core surrounded by phloem.

Amphivasal vascular bundle A vascular bundle with a core of phloem surrounded by xylem.

Anaphase The stage in mitosis in which the sister chromatids migrate to opposite poles of the spindle.

Androecium The collective term for the stamens of a flower.

Angiosperm A seed plant in which the ovule(s) is enclosed within an ovary at the base of the carpel and with a stigmatic surface on which pollen is deposited.

Anther The pollen-bearing terminal region of a stamen.

Anticlinal Refers to the formation of a new cell wall at right angles to the surface of the organ or duct (cf., periclinal).

Antipodal cell One of the cells, usually three, occurring at the chalazal end of the embryo sac.

Apoplast The non-protoplasmic region of the plant comprising cell walls, the lumina of dead cells and intercellular spaces.

Arborescent species Woody plants which are either trees or shrubs at maturity.

Areole The smallest area of leaf mesophyll (especially in dicotyledons) which is completely invested by veins.

Auricle In grasses one of the two lateral flaps of tissue located at the junction of the lamina and leaf sheath.

Axil The angle between the stem and the adaxial insertion of the leaf; it normally bears an axillary bud.

Axial parenchyma The longitudinally orientated parenchyma cells of the secondary vascular tissue (cf., ray parenchyma).

Axile placentation In a syncarpous gynoecium the longitudinal arrangement of the placentae on the central axis of the ovary (cf., parietal placentation).

Basal meristem In the leaves of many monocotyledons, especially perennials, the base of the leaf remains meristematic and new leaf tissue continues to form from this source.

Bark The inner bark of a tree comprises the vascular cambium and youngest secondary phloem, whilst the outer bark corresponds to the rhytidome.

Bicollateral bundle A vascular strand in the shoot with phloem forming to the outside and inside of the xylem (cf., collateral bundle).

Bifacial leaf A dorsiventral leaf showing palisade mesophyll on the upper (adaxial) surface and spongy mesophyll on the lower surface (cf., isobilateral leaf).

Bisporic development The embryo sac which forms when one of the two derivatives of the first meiotic division degenerates but the other proceeds to the second division; both haploid nuclei thus formed undergo mitosis to form the mature embryo sac (cf., mono- and tetrasporic development).

Bryophytes A group of non-vascular land plants comprising the liverworts and mosses in which the gametophyte stage is dominant in contrast to vascular plants.

Bundle sheath Layer(s) of parenchyma or sclerenchyma cells enclosing a vein of the leaf.

Callose An amorphous polysaccharide common in the walls of sieve areas, of phloem. It also forms rapidly in wounded sieve tubes where it helps to seal the sieve pores, and callose is also synthesised in developing pollen tubes.

Callus An unorganised tissue mass, initially composed mainly of parenchyma, formed at the wounded surface of plant organs and protecting the plant from infection. Later a protective layer of cork may develop within it, whilst adventitious roots and shoots sometimes form from meristematic nodules in the callus.

Calyptrogen A distinct meristematic layer in the root apex, present in the grasses and some other roots, which gives rise to the root cap.

Cambium A lateral meristem which either forms secondary xylem and phloem (vascular cambium) or cork (cork cambium).

Carbohydrate A general term for substances composed of carbon, hydrogen and oxygen and having the general chemical formula of $C_nH_{2n}O_n$

Carpel A component of the gynoecium which may be free or fused to other carpels and bears ovule(s) in its ovary.

Casparian band A continuous impermeable layer, composed of lignin and suberin, located in the radial and transverse (anticlinal) primary walls of the endodermis in roots and some stems.

Cavitation The formation of a bubble in the water column within a vessel or tracheid leading to the loss of their water conducting capacity.

Cell plate The partition formed from fused golgi vesicles, which separates the two nuclei at the end of mitosis and spreads centrifugally to divide the mother cell. The pectin interior of the plate constitutes the middle lamella and subsequently primary wall is secreted on either side of it.

Cell wall The protoplast of a plant cell is normally surrounded by a fairly rigid primary wall composed of a fibrillar cellulosic framework linked to amorphous polysaccharides and proteins. A secondary wall may be deposited internally, this usually contains a higher proportion of cellulose and is often lignified.

Cellulose A polysaccharide consisting of glucose molecules linked into long unbranched drains of up to 15,000 monomers. The chains are laterally hydrogen bonded to form microfibrils up to several micrometres long and 3–8 nanometres wide.

Central mother cells The terminal zone of the shoot apex from which the subjacent apical tissue is derived.

Centrifugal growth Development from the centre towards the outside as in the growth of the cell plate or differentiation of the primary xylem in the shoot.

Centripetal growth Development from the outside towards the centre as in the differentiation of the primary xylem in the root.

Chalaza Region of the ovule where the base of the nucellus is attached to the funiculus.

Chimaera A plant or organ composed of tissues of several genotypes; as in a shoot apex in which a mutation in an initial leads to its derivative tissues being incapable of developing chloroplasts and the leaf appearing variegated.

Chitin A polymer formed from a modified sugar molecule; it is the main skeletal material in the cell wall of fungi and also occurs in insect and crustacean outer skeletons.

Chloroplast A plastid concerned with photosynthesis. The internal chlorophyll bearing membranes are very extensive and normally arranged into a complex series of stacked cisternae forming numerous grana which interconnect by stromal membranes. The surrounding stroma contains the enzymes for carbon fixation.

Chromatid The half chromosome (joined to its partner by the centromere) visible during early mitosis and also in meiosis.

Chromatin See eu- and heterochromatin.

Chromosome A body within the nucleus bearing genes arranged linearly. The chromosomes are normally decondensed in the interphase nucleus and not distinguishable by the light microscope, but during nuclear division they form visible thread-like bodies.

Cisternae A flattened membranous compartment bounded by a single membrane as in the endoplasmic reticulum and golgi body.

Cleistogamy Self-pollination within an unopened flower bud.

Coated vesicle A small cytoplasmic vesicle coated with clathrin and apparently pinched off from the plasmalemma.

Coenocyte A large cell containing several to many nuclei, usually resulting from mitosis unaccompanied by cytokinesis.

Coleoptile A sheath which encloses the embryonic shoot in grass grains. During germination it forms a protective channel for the elongating shoot.

Coleorhiza A sheath enveloping the radicle of grass and certain other monocotyledonous embryos.

Collateral bundle A strand of vascular tissue with

xylem and phloem on the same radius and with the latter usually lying nearest the epidermis

Collenchyma A living supportive tissue consisting of elongate cells with unevenly thickened, non-lignified walls. It is common in the peripheral regions of the young shoot.

Companion cell A specialised parenchyma cell with extensive plasmodesmatal connections to a sieve tube member; both cells are derived from a common mother cell. The densely cytoplasmic companion cells apparently control the functioning of the enucleate sieve tubes.

Complementary tissue The cork cells underlying lenticels have abundant intercellular spaces, unlike the impermeable cork elsewhere, which allow aeration of the internal living tissues.

Compression wood In gymnosperms the wood formed in the lower sides of branches which is dense and heavily lignified (cf., tension wood).

Conifer A cone-bearing tree belonging to the largest division of the gymnosperms; common members are pines, firs and larches.

Corm A short, vertically-orientated, swollen underground stem storing food and allowing the plant to perrenate.

Contractile root The contraction of such roots keeps the shoot at a constant level to the soil surface; they are common in rhizomes and underground stems.

Cork A non-living protective layer composed of radially aligned cells with suberised and impermeable walls. It replaces the epidermis in many woody stems and roots and is formed centrifugally from a cork cambium (phellogen).

Cork cambium A lateral meristem arising in woody stems or roots which divides periclinally to give cork (phellem) centrifugally and sometimes to parenchyma tissue (phelloderm) centripetally.

Corpus The inner region of the shoot apex of flowering plants which is covered by the tunica. The cells of the corpus can divide in any plane whereas the tunica cells divide only in an anticlinal plane.

Cotyledon A first-formed leaf on the embryo; in monocotyledons only a single cotyledon is present but in dicotyledons two are present. In many plants the cotyledons are greatly modified food storage organs and do not develop into normal foliage leaves on germination.

Crista (pl. cristae) The tubular internal extension into the stroma of the inner membrane of the mitochondrial envelope.

Cuticle A layer of fatty material (cutin) covering and partially impregnating the outer epidermal walls of the shoot. The cuticle is thick and conspicuous in shoots of xerophytic plants where its waterproofing properties greatly impede water loss.

Cycad A primitive group of cone-bearing gymnosperms, with large palm-like leaves (as in *Cycas*), confined to the tropics and sub-tropics.

Cytokinesis The division of a cell into two by a cell wall after nuclear division.

Cytoplasm The living components within the cell wall, except for the nucleus and vacuoles, constitute the cytoplasm.

Cytosol The liquid phase of the cell in which the cytoplasmic organelles are suspended.

Deciduous plants Trees and shrubs which lose their leaves at the end of the growing season.

Dedifferentiation The cytological and biochemical changes accompanying the division of totipotent parenchyma cells and their reversion to small, densely cytoplasmic, meristematic cells. These events often accompany wounding to the plant and adventitious organogenesis *in vitro* and *in vivo*.

Dehiscent fruit A fruit which when ripe splits open to release the seeds

Dermal tissue The external covering tissue of the plant comprising the epidermis or periderm.

Desmotubule The fine tubular thread which traverses the plasmodesma and is linked at either end to the endoplasmic reticulum of the associated protoplasts.

Diarch Refers to roots in which two protoxylem poles are visible in transverse section.

Dichotomy The division of an apical meristem into two, usually equal, components. Also the venation of leaves in which the main veins divide into equal components.

Dictyosome A cellular organelle (also termed Golgi body) consisting of a stack of plate-like membranous cisternae. Vesicles, or sometimes cisternae, become detached from this body and transport carbohydrates and glycoproteins to the plasmalemma where they are voided into the cell wall.

Differentiation The biochemical and structural changes occurring in an individual cell tissue, organ or the whole plant during its growth and development from an immature to a mature form.

Diffuse porous Dicotyledonous secondary xylem in which the vessels occur fairly uniformly throughout one season's growth of wood.

Diffuse secondary growth In certain palms and other monocotyledons the basal regions of the trunk may widen due to diffuse division of the parenchymatous ground tissue and bundle sheath parenchyma.

Distal Farthest from the point of attachment or origin (cf., proximal).

Double fertilisation An event unique to angiosperms in

which one male gamete fertilises the egg to give a diploid zygote whilst the other sperm fertilises the (diploid) polar nucleus to give a triploid primary endosperm nucleus.

Ectomycorrhizal association A fungus associated with the roots of certain trees (pine, beech, birch). The hyphae form a dense covering to the roots and also ramify between the outer cortical cells.

Egg The haploid female cell which is fertilised by a sperm to give rise to a diploid zygote.

Embryo The young plant present in the seed.

Embryo sac The female gametophyte of flowering plants retained within the nucellus of the ovule. The sac typically shows eight haploid nuclei contained within an egg consisting of two synergid cells at the micropylar pole, three antipodal cells at the chalazal end and one binucleate central cell.

Endarch Primary xylem in the stem in which the protoxylem lies nearest the centre of the stem and the metaxylem towards the outside.

Endocarp The inner layer of the fruit wall which is often sclerified.

Endodermis A layer of cells surrounding the vascular system of roots and some stems. Initially each endodermal cell shows a continuous Casparian band of ligno-suberin within the anticlinal walls but later more extensive layers of thickening may be deposited. The impermeable deposit prevents apoplastic movement of water and solutes across the endodermis so that only symplastic transport is possible.

Endomycorrhizal association A symbiotic association between a fungus and the roots of many plants. The mycelium ramifies internally and often invades the root cell where it forms a vesicular or branched structure.

Endoplasmic reticulum (ER) A cisternal or tubular membranous system bounded by a single membrane, which ramifies through the cytoplasm and shows connections to the outer nuclear membrane and the desmotubules. The rough endoplasmic reticulum (RER) is coated with ribosomes and concerned with protein synthesis and the rarer smooth form (SER) with lipid metabolism.

Endosperm The nutritive tissue (usually triploid) resulting from the fusion of the sperm with the central cell in angiosperms. The endosperm helps nourish the developing embryo and is frequently present as a food reserve in mature monocotyledonous seeds and some dicotyledons.

Endothecium A hypodermal layer of the anther wall often containing wall thickenings concerned with dehiscence of the ripe anther.

Epidermis The outermost layer of the primary shoot and root. It is normally a discrete single layer but periclinal divisions within it rarely give rise to a multiple epidermis (e.g. *Ficus* leaf, root velamen of epiphytic orchids).

Epiphyte A plant, growing in a suitable niche on the surface of another plant, which is neither symbiotic nor a parasite. Especially in tropical regions, the trunks of many trees are covered by various epiphytic flowering plants, ferns and bryophytes.

Establishment growth The early period of development in palms and certain other plants, during which the trunk is attaining the thickness characteristic of the mature individual. Subsequently, elongation of the trunk occurs.

Etioplast A plastid characteristic of potentially green tissues but which develop in the dark. On exposure to light the internal membranes rapidly form a granal system and a green chloroplast results.

Euchromatin The chromatin (DNA and histone) which appears as lightly-stained regions of the interphase nucleus in material viewed in the light or electron microscope (cf. heterochromatin).

Exarch Primary xylem in which the protoxylem differentiates towards the outside of the organ and the metaxylem towards the centre, as in seed plant roots.

Exocarp The external layer of a fruit wall (syn. epicarp).

Exine The outermost layer of a pollen grain or spore which is very resistant to decay due to the deposition of sporopollenin within it (cf. intine).

Exodermis In some roots a layer of outer cortical tissue becomes impermeable due to the deposition of ligno-suberin in their cell walls.

Fascicular vascular cambium This originates within the procambial strands of the shoot in dicotyledons and gymnosperms and lies between the xylem and phloem.

Ferns Perennial, mainly herbaceous, vascular plants which are non-seed bearing. The dominant sporophyte generation bears sporangia where haploid spores are produced to give rise to small, free-living and autotrophic gametophytes.

Fibre An elongated and usually tapered sclerenchyma cell with thick, usually lignified, second walls. It is usually dead at maturity.

Filiform apparatus Within the embryo sac of some species the synergids develop labyrinthine transfer walls where they contact the egg and nucellus.

Fixation process The killing and preservation of the cellular structure of biological tissues, so that material can be examined under the microscope in a nearly life-like form.

Flank meristem In some angiosperms the cells of the shoot apex show variations in the density of their staining: in such situations the more densely-staining marginal tissue is designated as flank meristem.

Freely-forming walls In some situations within the plant, mitosis is not immediately followed by cell division so that coenocytic cytoplasm is formed. Walls may subsequently develop (as in the endosperm of angiosperms) but such walls are tortuous and cellularisation is apparently haphazard.

Fret The cisternal or tubular membranous connection extending through the chloroplast stroma from one granum to another.

Freeze-fractured material Tissue which has been rapidly frozen, so that ice crystals are normally absent, and then broken across under high vacuum. The fractured surface is shadowed with platinum, followed by a stabilising layer of carbon. The resulting replica of the surface is examined under the electron microscope.

Fungus A plant-like, spore-bearing organism, which lacks chloroplasts. It has heterotrophic nutrition and is either a parasite or saprophyte.

Fruit Confined to angiosperms; the structure which develops from the enlarged ovary and contains the seeds. In some species parts additional to the ovary are incorporated in the fruit, as in the apple, stawberry and pineapple.

Funiculus The stalk connecting the ovule to the placenta of the ovary.

Fusiform initial An elongate, tapering cell located in the vascular cambium from which axial elements of the secondary vascular tissues originate (cf., ray initial).

Gametophyte The haploid phase of the life cycle; in bryophytes the gametophyte is dominant (cf., sporophyte).

Generative cell In pollen the cell which divides to form two male gametes.

Glyoxysome A single membrane-bounded organelle involved in glyoxylic acid metabolism; abundant during germination of seeds containin lipid stores.

Golgi body An alternative term for dictyosome.

Granum (pl. grana) In a chloroplast a stack of discoidal cisternae, each bounded by a single membrane, in which the chlorophyll and carotenoid molecules are located.

Ground tissue The tissues of the plant body excluding the vascular and dermal systems.

Growth ring A layer of secondary xylem or phloem visible in a cross section of a woody stem or root.

Guard cells A pair of specialised epidermal cells in the shoot which border the stomatal pore; changes in their turgor causes the opening and closing of the stoma.

Gymnosperms Seed plants in which the ovules are not enclosed within an ovary (cf., angiosperm) as in *Pinus*, *Ginkgo* and *Cycas*.

Gynoecium The collective name for the carpels of a flower.

Halophyte A plant growing in saline conditions and often showing a succulent habit.

Haploid A plant having a single complete set of chromosomes (cf., diploid); the normal condition in the gametophytic stage of the life cycle.

Hardwood The wood of dicotyledonous trees (cf., softwood) which contains numerous thick-walled fibres in addition to tracheary elements.

Hastula A flap of tissue occurring in fan-leaved palms at the junction of the petiole and lamina.

Haustorium A penetrating and absorptive structure; for example in parasitic flowering plants (mistletoe and dodder) modified roots tap nutrients from the host plant.

Heartwood The darker-coloured central wood of a tree in which the tracheary elements are non-conducting and plugged with resins and tyloses whilst the parenchyma cells are dead.

Hemicellulose A group of polysaccharides of the plant cell wall composed of several different simple sugars in various combinations and not forming microfibrils (cf., cellulose).

Heterochromatin The densely-staining chromatin (DNA and histone) visible by light and electron microscopy in the interphase nucleus (cf., euchromatin).

Heterosporous Refers to all seed plants and a few lower vascular plants (*Selaginella*) in which the plant produces both mega- and microspores (cf., homosporous).

Hilum The scar on a seed showing the original attachment of the funiculus. Also, the centre of a starch grain around which layers of starch are successively deposited.

Homosporous Refers to plants in which all the spores produced are uniform in size and shape as in bryophytes and most ferns (cf. heterosporous).

Hypha A thread or filament of a fungus.

Hypocotyl The portion of the embryo or seedling, lying between the root and the insertion of the cotyledon(s).

Hypodermis Layer(s) of cells within the epidermis which is histologically distinct from the other ground tissue.

Indehiscent fruit A fruit which when mature does not rupture or open to release the seeds (cf., dehiscent fruit).

Inferior ovary An ovary lying beneath the level at which the other floral parts are inserted onto the receptacle (cf., superior ovary).

Internode A region of the stem located between successive leaves (cf., node).

Integument In seed plants the outer sterile jacket(s) of the ovule enclosing the nucellus except at the micropyle.

Intercellular space A gas space which forms between adjacent cells either by the breakdown of the middle lamellae or of an intervening cell.

Interphase nucleus The nucleus in the period between mitotic or meiotic division and in which discrete chromosomes are not discernable.

Intine The inner cellusosic wall layer of a pollen grain or spore (cf., exine).

Intrusive growth Elongation of a cell in which its growing tips intrude between the middle lamellae of adjacent cells, as in some fibres.

Isobilateral leaf A leaf in which palisade mesophyll occurs both ad- and abaxially (cf., bifacial leaf).

Kinetochore A specialised region of the chromosome at which the two chromatids are joined and from which microtubules originate to form part of the mitotic spindle.

Kranz anatomy The radial arrangement of the mesophyll cells around each bundle sheath in the leaf which is characterisitic of plants with C4 photosynthesis.

Lamina The blade of a leaf.

Lateral root A root arising from another root (cf., adventitious root).

Laticifer A secretory cell, or series of interconnected cells, containing the milky fluid latex.

Leaf An organ arising as a lateral swelling on the shoot apex and which typically bears a bud in its axil. The leaf is in vascular continuity with the stem and is normally the major photosynthetic region of the plant.

Leaflet In a compound leaf a series of individual leaflets arise from the axis of the leaf but do not subtend buds.

Leaf sheath In some leaves their basal portions invest the stem to form a distinct sheath, as in many monocotyledons.

Lenticels Pores in the bark formed, in contrast to the adjacent compact cork, from rounded cells with intercellular spaces which allow oxygen to diffuse into the plant.

Leucoplast A colourless plastid with little starch, such as is present in the leaf epidermis.

Liane A climbing plant with a long woody stem, especially prevalent in tropical forests.

Lichen A symbiotic association between an alga and fungus living together.

Lignin A complex substance containing various phenolics which is deposited in the cellulose walls of the sclerenchyma and tracheary elements; it increases their strength and renders the walls impermeable to water.

Ligno-suberin A complex of lignin and suberin deposited in the walls of the endodermis and exodermis of the root.

Ligule In grasses a membranous projection from the adaxial leaf surface at the base of the lamina.

Lipid A group of fats and fat-like compounds which are soluble in certain organic solvents but not in water.

Liverwort A small bryophytic plant (cf., moss); the often thallosic green plant represents the dominant gametophyte in contrast to vascular plants.

Lumen The central channel or space, formerly occupied by the protoplast, in dead sclerenchyma and tracheary elements.

Megaspore A large haploid spore in seed plants and certain ferns and their allies, which develops into the female gametophyte or embryo sac.

Megasporangium A sporangium in which each diploid megaspore mother cell gives rise by meiosis to four haploid megaspores (cf., microsporangium).

Meristem A tissue primarily concerned with growth and division in an organised manner, as in the shoot and root apex, vascular and cork cambia.

Meiosis A sequence of two specialised nuclear divisions of a diploid cell resulting in the formation of four haploid cells.

Mesocarp The middle, often fleshy layer of the fruit wall.

Mesophyll The photosynthetic parenchyma of a leaf, frequently divided into cylindrical palisade cells and irregular spongy mesophyll.

Metabolism The process in which nutritive material is synthesised into protoplasm and cell wall or in which the latter are broken down into simpler substances.

Metaphase A stage in mitosis during which the kinetochores of the chromosomes all lie at the equator of the mitotic spindle.

Metaphloem The last-formed region of the primary phloem which matures after the organ has ceased to elongate.

Metaxylem The last-formed region of the primary xylem which matures after the organ has ceased to elongate; its tracheary elements are usually scalariform, reticulate or pitted.

Microbody An organelle demarcated by a single membrane and containing various non-hydrolytic enzymes (cf., glyoxysome and peroxysome).

Microfibril A series of cellulose molecules linked together by hydrogen bonding to form a fibril up to several micrometres long; microfibrils form the skeletal framework of the cell wall.

Microfilament A proteinaceous and filamentous component of the cytoskeleton of some plant cells; it is ca. 7 nanometres wide and narrower than a microtuble.

Micropyle A narrow pore in the integument(s) at the apex of the ovule, via which the pollen tube frequently penetrates the embryo sac.

Micrometre (μm) A unit of length representing one thousandth of a millimetre.

Microspore A haploid spore which develops into the male gametophyte; the pollen grain of seed plants.

Microsporangium This gives rise to numerous microspores (cf., megasporangium).

Microtubule A hollow proteinaceous tubule ca. 25 nanometres wide (cf., microfilament). Microtubules form the major component of the plant cytoskeleton and are located in the peripheral cytoplasm of non-dividing cells and also form the spindle fibres of dividing nuclei.

Middle lamella A layer of mainly pectic materials, derived from the cell plate, which cements together the primary walls of adjacent cells.

Midrib In many simple leaves a single prominent longitudinal rib extends the length of the leaf, consisting of a large vascular bundle(s) and sheath.

Mitochondrion (pl. mitochondria) An organelle delimited by an envelope whose inner membrane is involuted into tubules or cristae; responsible for aerobic respiration.

Mitosis The division of a diploid nucleus into two diploid daughter nuclei.

Mitotic spindle The fibrillar structure formed early in mitosis, whose 'fibres' are visible under the light microscope, and consist of fasciated microtubules. It is concerned in the segregation of the two chromatids of each chromosome to a different daughter nucleus.

Monocotyledon One of the two groups comprising the flowering plants; the monocotyledonous embryo has a single cotyledon (cf., dicotyledon). A number of other features (flower parts in threes, normally absence of secondary thickening, scattered arrangement of vascular bundles in the shoot, etc) also characterise a monocotyledon.

Monosporic development This is the typical situation in the formation of an embryo sac of a flowering plant in which only one of the four haploid cells, derived from the meiosis of the megaspore mother cell, undergoes development whilst the others abort.

Morphology The external form and development of the plant.

Moss A small, leafy bryophytic plant (cf., liverwort); the gametophyte generation is dominant in contrast to the situation in vascular plants.

Mucigel The growing root tip secretes mucilage from its cap cells which lubricates the passage of the root between the soil particles and may be important in nourishing beneficial soil microorganisms.

Multiple epidermis A several-layered tissue derived from the protodem by both periclinal and anticlinal divisions; only the outer layer forms a typical epidermis.

Multiple fruit This is derived from the ovaries of several to many individual flowers as in pineapple (cf., aggregate fruit).

Mycorrhiza The symbiotic association between the roots of many plants and soil fungi (see ecto- and endomycorrhiza).

Nectary A multicellular gland secreting nectar; present as floral nectaries in many plants but also occurring as extra-floral nectaries on the vegetative plant.

Nectar A fluid secreted by a nectary; the liquid is rich in sugars and other organic substance.

Nexine The inner layer of the exine in the wall of a pollen grain.

Node The region of the stem from which a leaf or leaves arise (cf., internode).

Nucellus The inner region of the ovule surrounding the embryo sac; considered to be homologous with the megasporangium.

Nuclear envelope The double membrane enclosing the nucleoplasm; the envelope is frequently penetrated by pores and the outer membrane is linked to the endoplasmic reticulum.

Nuclear pore The outer and inner nuclear membranes are often joined to form pores in the nuclear envelope.

Nucleolus A densely-staining granular body, commonly spherical, which occurs in the interphase nucleus; it is composed of RNA and protein and is the site of ribosome synthesis.

Nucleus A large organelle bounded by a double membrane and containing the chromosomes, nucleolus and nucleoplasm; commonly only a single nucleus is present per cell but some cells are coenocytic.

Organelle A cytoplasmic body with a specialised function.

Osteosclereid An elongate sclereid with enlarged ends.

Ovary The basal region of a carpel which contains one to many ovules; after fertilisation the ovary enlarges and differentiates to form the fruit.

Ovule In seed plants the female gametophyte (embro sac) is enclosed within the nucellus and integument(s). These, together with the funiculus, comprise the ovule which later develops into the seed.

P-protein The proteinaceous occurring in sieve tubes which in damaged tissue form a plug blocking the sieve plate.

Palisade mesophyll The parenchymatous tissue whose cylindrical cells lie with their long axes perpendicular to the epidermis; this compact photosynthetic layer occurs on the adaxial surface of bifacial leaves.

Parenchyma cell An unspecialised, highly vacuolated cell with typically only a primary wall of uniform thickness; it occurs as extensive regions of tissue in the pith, cortex and mesophyll of the plant body.

Parietal placentation Occurs in an ovary in which the ovules are attached to peripheral placenta(e).

Parthenocarpy Development of a fruit without fertilisation of the ovules.

Perennial A plant whose vegetative body persists for many years.

Perforation plate The end wall of a vessel element, either a single large hole (simple plate), commonly lying in a transverse wall, or several pores forming a compound plate on an oblique end wall.

Perianth The collective name for the sterile outer parts of a flower which are often differentiated into the outer sepals and inner petals.

Pericarp A synonym for fruit wall.

Periclinal Refers to a cell wall forming parallel to the surface of an organ (cf., anticlinal).

Pericycle In a root the tissue, usually parenchymatous, lying between the endodermis and vascular tisses.

Periderm The secondary protective tissues (bark) replacing the epidermis; it comprises the phellem, phellogen and phelloderm, plus any cortex or phloem lying internal to a deeply situated cork cambium.

Perinuclear space The region lying between the two membranes of the nuclear envelope.

Periplasmodium The coenocytic mass, formed from fused tapetal protoplasts, which occurs around the developing pollen grains in some anthers.

Perisperm A nutritive storage tissue formed from the nucellus present in the seeds of several dicotyledonous families.

Peroxysome A single membrane-bounded microbody lying adjacent to a chloroplast and involved in the metabolism of glycolic acid associated with photosynthesis.

Petal An inner perianth member which is distinct in form from a sepal.

Petiole The narrow stalk which attaches the leaves of many plants to the stem.

Phellem The non-living outer layer (cork) of the periderm which is impermeable due to the deposition of suberin within the walls of its constituent cells; formed from the phellogen.

Phelloderm A parenchymatous tissue which is formed centripetally from the phellogen of some plants.

Phellogen A lateral meristem (cork cambium) which by regular periclinal divisions forms phelloderm to its exterior; in some plants a phelloderm is also formed to the interior.

Phloem The main food transporting tissue of vascular plants; consisting of the conducting sieve elements, various types of parenchyma and sclerenchyma.

Phragmosome The layer of cytoplasm stretching across a vacuolated cell in which the nucleus is situated during division, and which demarcates the plane of newly-forming cell plate.

Phylloclade Interpreted as a petiole which has become flattened and replaces the lamina of a leaf.

Phyllotaxy The pattern in which the leaves are arranged on the stem.

Pit A region of the cell wall in which the primary wall remains uncovered by the deposition of secondary wall; the recess in the wall may be of uniform width (simple pit) or the pit is bordered at its outer margin.

Pitfield A thin region of primary wall with numerous plasmodesmata; if secondary wall is later deposited, a pit develops over this region.

Pith The central ground tissue (usually parenchymatous) of the stem and some roots.

Placenta The regions of the ovary to which the ovules are attached.

Plasmalemma The single membrane which demarcates the cell protoplast from the externally lying wall; the plasmalemmae of contiguous cells are in continuity via their plasmodesmata.

Plasmodesma (pl. plasmodesmata) A pore in the cell wall linking adjacent protoplasts; it is lined by plasmalemma and contains an axial desmotubule linked to endoplasmic reticulum at either end.

Plastid The generic name for a varied group of organelles, bounded by a double membrane, which are derived from a proplastid; common examples are chloroplasts and amyloplasts.

Plastoglobulus (pl. plastoglobuli) A small densely-staining lipidic vesicle occurring within the stroma of chloroplasts and other plastids.

Plicate mesophyll cell A parenchyma cell whose primary walls are enfolded into the protoplast, as in pine leaves.

Plumule The region lying above the cotyledon(s) in the embryo and which forms the young shoot in the seedling (cf., radicle).

Pneumatophore A negatively geotropic root projecting from the substratum; produced by trees living in swamp conditions and serving for aeration of the underground root system.

Polar nucleus In the ovule of a flowering plant one of the (normally) two nuclei occurring in the central cell; the endosperm results from the fusion by a male gamete with these nuclei.

Pollen grain In seed plants the term used for microspore.

Pollen tube The tube developing from a germinated pollen grain, in which the male gametes are transported to the embryo sac.

Polyarch The roots of monocotyledons have numerous protoxylem poles and are termed polyarch.

Polyploid Referring to a plant or cell possessing a multiple of the normal diploid set of chromosomes, e.g.

a tetraploid has a double set.

Polysaccharide A carbohydrate composed of many monosaccharides linked together in a chain, as cellulose and starch.

Polysome A complex of ribosomes concerned with protein synthesis.

Primary cell wall The wall formed by the protoplast up to the end of expansion growth; the cellulose microfibrils are often randomly orientated and are less abundant than in the secondary wall.

Primary thickening meristem This occurs in the subapical region of plants with greatly thickened primary stems; divisions throughout the incipient cortex, procambium and pith lead to rapid radial growth of the axis, as in cycads and monocotyledons.

Procambium A meristematic tissue arising directly from the apical meristem; in the primary plant body it differentiates into the primary vascular tissues; in dicotyledons and gymnosperms it also forms the fascicular cambium.

Proembryo The embryo before the onset of organ and tissue differentiation.

Prolamellar body The star-shaped complex of membranous tubules occurring in an etioplast; on exposure to light this body is transformed into grana and frets.

Proleptic growth Rhythmic growth of a perennial plant (cf., sylleptic).

Prop root An adventitious root formed on the stem above the soil surface and helping to anchor the plant.

Prophase The early stage of nuclear division; characterised by the appearance of the chromosomes, the breakdown of the nuclear envelope and the development of the spindle apparatus.

Proplastid A small and undifferentiated plastid occurring in meristematic tissues; the progenator of all other plastid types.

Protein A large and complex molecule composed of various amino acids.

Protoderm Meristematic tissue which gives rise to the epidermis.

Protophloem The first phloem to differentiate from the procambium and usually consists of sieve elements only. These are short-lived and usually crushed in the developing shoot (cf., metaphloem).

Protoplast The protoplasm confined within the walls of an individual cell.

Protoxylem The first xylem to differentiate from the procambium, usually consisting of annular or spirally-thickened tracheary elements (cf., metaxylem).

Proximal Nearest the point of attachment or origin (cf., distal).

Pulvinus A joint-like thickening of the leaf petiole (or of a petiolule) in which the central vascular strand is surrounded by a broad expanse of parenchyma. Loss of turgor in this tissue causes the leaf to droop.

Quiescent centre The terminal region of the root apex in which cell divisions are absent, or occur very infrequently relative to the adjacent meristematic cells.

Radicle The embryonic root situated beneath the hypocotyl in the seed and forming the main root of the seedling (cf., plumule).

Ray A panel of parenchyma extending radially across the secondary vascular tissues; a ray is formed from an initial in the vascular cambium and is of variable width and height.

Ray initial A squat, semi-cuboidal, cell of the vascular cambium giving rise to the ray parenchyma of the secondary vascular tissues (cf., fusiform initial).

Receptacle The terminal region of the flower stalk to which the floral parts are attached.

Resin canal A long duct lined with epithelial cells which secrete the sticky resin common in conifers.

Rib meristem The sub-terminal axial region evident in some shoot apices; its derivatives divide predominantly transverse to the long axis of the young stem and give rise to the pith.

Rhizome An elongate horizontal stem growing beneath the soil; a common organ of perennation in monocotyledons.

Rhizosphere The region of soil immediately surrounding the root.

Rhytidome The outer bark inclusive of the periderm and any cortical and phloem tissues isolated from the functional phloem by a deep-sited phellogen.

Ribosome A small organelle composed of RNA and protein which is concerned with protein synthesis; ribosomes may be aggregated into polysomes.

Ring porous wood Secondary xylem with the vessels of the spring wood much wider and more numerous than in later wood; this pattern of vessel formation leads to rings being visible in transverse section.

Root A plant organ which is linked to the shoot and is typically subterranean; roots are primarily concerned with absorption of water and mineral salts, anchorage and nutrient storage.

Root cap A cap of cells enclosing the root apex.

Root hair A simple cylindrical bulge from an epidermal cell of the young root which extends laterally between the adjacent soil particles and extends the absorptive surface of the root.

Root pressure The water pressure in the xylem resulting from the active transport of mineral salts into the vascular cylinder by the endodermis, thus causing intake of water from the cortex.

Rosette A shoot with very short internodes but bearing

fully expanded leaves. Also a group of cellulose-synthesising enzymes located at the plasmalemma.

Scale leaf A non-foliage leaf often investing dormant buds or found in underground stems.

Schizogeny The separation of cells along their middle lamellae to form an intercellular space.

Sclereid A type of sclerenchyma cell characterised by its very thick lignified walls and numerous pits; the shape is variable but it is generally much shorter than a fibre.

Sclerenchyma A supporting tissue whose cells are commonly dead at maturity and possess thick, lignified secondary walls, as in fibres and sclereids.

Scutellum The highly modified cotyledon present in grasses which supplies nutrients from the endosperm to the germinating embryo.

Secondary cell wall The wall formed by the protoplast at the end of expansion growth; the cellulose microfibrils are closely crowded and, in any one layer, lie parallel to each other (cf., primary wall). Secondary walls often become lignified, as in sclerenchyma and tracheary elements.

Secondary thickening This occurs in gymnosperms and most dicotyledons and some anomalous monocotyledons. The stem and root increase in diameter due to the formation of secondary vascular tissues by the vascular cambium (or by the secondary thickening meristem), whilst the epidermis is normally replaced by cork formed from the cork cambium.

Secondary thickening meristem In some arborescent monocotyledons (e.g. *Dracaena*, *Cordyline*) an anomalous form of secondary thickening occurs from a meristem which arises in the outer cortex and cuts off discrete vascular bundles, plus parenchyma, centripetally.

Seed The structure which develops from the fertilised ovule; it contains the embryo and a food supply to support early seedling growth.

Sepal An outer perianth member which is distinct in form from a petal.

Septate fibre A fibre with thin cross walls which develop after the longitudinal walls have become thickened.

Sexine The outermost region of the pollen grains ectine (cf., nexine).

Shoot The non-root region of the plant; it is usually aerial and composed of the stem bearing numerous photosynthetic foliage leaves (cf., root).

Sieve area Modified pit fields in the side and oblique end walls of sieve cells or tubes; the plasmodesmata have been transformed into narrow sieve pores and lateral translocation probably occurs via them.

Sieve cell The enucleate translocating element in gymnosperms and lower vascular plants, possessing sieve areas on all walls.

Sieve plates The transverse or somewhat oblique walls occurring in sieve tubes; it contains either a single series of large pores or is compound with several series of pores (cf., sieve area).

Sieve pore The hole in a sieve area or plate through which cytoplasmic continuity occurs from one sieve element to another; the wall surrounding the pore is commonly impregnated with callose.

Sieve tube An elongate element comprising several to many enucleate cells interconnected via the sieve plates (former cross walls). Sieve tubes are confined to angiosperms (cf., sieve cell).

Simple fruit Formed from the single ovary of an individual flower.

Softwood The wood of a conifer which generally lacks thick-walled, lignified fibres; it is therefore easier to saw than most hardwoods.

Spongy mesophyll Very irregular green parenchyma cells with large intercellular spaces between them; in bifacial leaves this tissue occurs abaxially (cf., palisade mesophyll).

Sporangium A structure in which spores are produced; in most ferns the spores are uniform in size but in seed plants different sized spores are produced in mega- and microsporangia.

Spores Haploid cells formed as derivatives of the meiotic division of a diploid spore mother cell within a sporangium. Each spore germinates to form the gametophyte.

Sporophyte The diploid phase of the life cycle; in vascular plants the sporophyte is dominant (cf., gametophyte).

Sporopollenin The substance composing the exine of pollen grains; it is formed from cyclic alcohols and is highly resistant to microbial decay.

Starch The chief food storage polysaccharide of plants composed of several hundred hexose sugars; it is insoluble and accumulates within the stroma of various plastids.

Starch sheath In many primary dicotyledonous stems the inner layer of the cortical parenchyma forms a sheath with rich deposits of starch in its cells.

Stamen The male organ of the flower composed of the terminal anther bearing pollen and the basal sterile filament.

Stigma The receptive zone of a carpel at the tip of the style, upon which pollen is deposited and germinates.

Stipules Projections of tissue on either side of the base of the leaf, which in dicotyledons are sometimes large and vasculated.

Stoma (pl. stomata) A complex consisting of a pore in the shoot epidermis which is surrounded by two specialised guard cells; their turgidity causes the opening and closing of the stomatal pore and thus controls gaseous exchange with the external atmosphere.

Stone cell A small, thick-walled and more-or-less isodiametric sclereid.

Stroma The non-membranous ground substance of a plastid.

Style The region of a carpel lying between the stigma and ovary.

Suberin A fatty, hydrophobic deposit in the cell walls of cork, associated with lignin in the thickened walls of the root endodermis and exodermis.

Subsidiary cells In some plants these occur adjacent to the guard cells of a stoma; subsidiary cells are morphologically distinct from the general epidermal cells.

Succulent A plant with fleshy leaves and stems containing many large, water-storing parenchyma cells.

Superior ovary An ovary which is inserted into the receptacle above the level of the other floral parts.

Suspensor A multicellular structure, usually filamentous, which is anchored at one end near the micropyle of the ovule and at the other to the radicle pole of the embryo; extension of the suspensor pushes the growing embryo into the endosperm.

Sylleptic growth Continuous growth of a perennial plant without rest phases (cf., proleptic growth).

Symbiosis The mutually beneficially association of two different kinds of living organisms (e.g. lichens, nitrogen fixing root nodules).

Symplast The combined protoplasts of the plant body; these are all linked by their plasmodesmata (cf., apoplast).

Synergids In the embryo sac of flowering plants the two cells adjacent to the egg, which apparently play an essential role in the transmission of the male gametes to the egg and polar nuclei.

Tapetum The layer of nutritive cells lining the pollen sac; the tissue is absorbed as the pollen grains mature.

Telophase The last stage of mitosis in which the two daughter nuclei are reorganised at the poles of the mitotic spindle.

Tension wood Forms on the upper sides of branches in arborescent dicotyledons and is characterised by the occurrence of numerous fibres whose walls are non-lignified and highy hydrated (cf. compression wood).

Tap root The main root of many dicotyledons which is directly derived from the persistent radicle.

Testa The investing layer of a seed formed from the modified integuments of the ovule.

Tetrasporic embryo sac An embryo sac which develops when all four haploid nuclei, formed by the meiosis of the megaspore mother cell, survive and contribute to the mature embryo sac.

Thallus The non-leafy, dorsiventrally flattened gametophyte of many liverworts.

Thylakoids The photosynthetic internal membranes of a chloroplast consisting of grana and frets.

Tissue A group of cells forming a discrete functional unit; in simple tissues the cells are all alike, whereas in a complex tissue varied cell types occur.

Tonoplast The single membrane which encloses a vacuole.

Totipotent A differentiated plant cell or tissue (e.g. parenchyma) which retains all of the genetic material present in the embryo; due to wounding or hormonal influence such cells may undergo dedifferentiation, regain their meristematic capacity, and give rise to adventitious organs and embryos.

Torus The bi-convex thickened disc forming the central part of the primary wall of a bordered pit in conifers.

Tracheary element A collective term for the vessels and tracheids of the xylem. These dead, water-conducting elements show various patterns of secondary wall thickening (annular, spiral, scalariform, reticulate and pitted).

Tracheid An elongated imperforate tracheary element with various patterns of secondary wall deposition (cf., tracheary element).

Transfer cell This shows labyrinthine ingrowth of its walls which greatly increase the surface area of plasmalemma; these cells function in the large-scale transport of solutes over short distances.

Transfusion tracheids Specialised xylem, confined to gymnosperm leaves, in which the tracheids are short and with non-tapering ends.

Translocation The movement of sugars and other organic substances throughout the vascular plant body via the sieve elements of the phloem.

Transmitting tissue The specialised tracts of tissue in the style through which the pollen tubes grow towards the ovary.

Transpiration The movement of water from the root to the shoot in the tracheary elements of the xylem and the subsequent loss of water vapour from the leaf surface via the stomata.

Trichome Any outgrowth from an epidermal cell; it may be unicellular or multicellular and is often glandular.

Tuber A much-enlarged underground stem or root from which the plant perennates.

Tunica The outermost layer(s) of the shoot apex in flowering plants characterised by divisions which are only anticlinal (cf., corpus).

Tylose An outgrowth of a xylem parenchyma cell which penetrates through the pit of a tracheary element (commonly a vessel) and enlargens within its lumen to block it.

Vacuole An organelle bounded by the tonoplast and containing a watery fluid. During differentiation the many small vacuoles of the meristematic cell become confluent and expand to form a large central vacuole.

Variegated leaf A leaf showing a distinct pattern of green and lighter or white regions in the leaf blade; commonly formed from a varigated chimaeral shoot apex.

Vascular bundle A strand of tissue composed of primary xylem and phloem (and in dicotyledons cambium) running lengthwise in the shoot.

Vascular cambium See cambium.

Vascular tissue A general term referring to both the xylem and phloem.

Vegetative cell In flowering plants this is the largest of the two cells in the young pollen grain.

Velamen The water-absorptive multiple epidermis in the roots of tropical epiphytic orchids and on some aroids.

Venation The pattern formed by the veins supplying the leaf blade.

Vessel A long, perforate tracheary element, formed from a longitudinal series of cells by the breakdown of their end wall (see perforation plates). The pattern of secondary wall deposition is variable (cf., tracheary element).

Wood The secondary xylem of arborescent dicotyledons and of gymnosperms.

Xeromorphic Morphological features typical of xerophytes.

Xerophyte A plant of dry regions with various xeromorphic features such as hairy and inrolled leaves, thick cuticles, sunken stomata, or spiny reduced leaves with succulent photosynthetic stems.

Xylem A complex tissue composed of the water conducting tracheary elements, sclerenchyma and parenchyma.

Zygomorphic flower A flower with bilateral symmetry.

Zygote The diploid cell resulting from the fertilisation of a haploid egg cell by a male gamete; subsequent division of the zygote leads to the formation of the pro-embryo.

Index